Springer Tracts in Modern Physics
Volume 157

Managing Editor: G. Höhler, Karlsruhe

Editors: J. Kühn, Karlsruhe
Th. Müller, Karlsruhe
R. D. Peccei, Los Angeles
F. Steiner, Ulm
J. Trümper, Garching
P. Wölfle, Karlsruhe

Honorary Editor: E. A. Niekisch, Jülich

Springer
Berlin
Heidelberg
New York
Barcelona
Hong Kong
London
Milan
Paris
Singapore
Tokyo

Springer Tracts in Modern Physics

Springer Tracts In Modern Physics provides comprehensive and critical reviews of topics of current interest in physics. The following fields are emphasized: elementary particle physics, solid-state physics, complex systems, and fundamental astrophysics.
Suitable reviews of other fields can also be accepted. The editors encourage prospective authors to correspond with them in advance of submitting an article. For reviews of topics belonging to the above mentioned fields, they should address the responsible editor, otherwise the managing editor.
See also http://www.springer.de/phys/books/stmp.html

Managing Editor

Gerhard Höhler

Institut für Theoretische Teilchenphysik
Universität Karlsruhe
Postfach 69 80
D-76128 Karlsruhe, Germany
Phone: +49 (7 21) 6 08 33 75
Fax: +49 (7 21) 37 07 26
Email: gerhard.hoehler@physik.uni-karlsruhe.de
http://www-ttp.physik.uni-karlsruhe.de/

Elementary Particle Physics, Editors

Johann H. Kühn

Institut für Theoretische Teilchenphysik
Universität Karlsruhe
Postfach 69 80
D-76128 Karlsruhe, Germany
Phone: +49 (7 21) 6 08 33 72
Fax: +49 (7 21) 37 07 26
Email: johann.kuehn@physik.uni-karlsruhe.de
http://www-ttp.physik.uni-karlsruhe.de/~jk

Thomas Müller

Institut für Experimentelle Kernphysik
Fakultät für Physik
Universität Karlsruhe
Postfach 69 80
D-76128 Karlsruhe, Germany
Phone: +49 (7 21) 6 08 35 24
Fax: +49 (7 21) 6 07 26 21
Email: thomas.muller@physik.uni-karlsruhe.de
http://www-ekp.physik.uni-karlsruhe.de

Roberto Peccei

Department of Physics
University of California, Los Angeles
405 Hilgard Avenue
Los Angeles, CA 90024-1547, USA
Phone: +1 310 825 1042
Fax: +1 310 825 9368
Email: peccei@physics.ucla.edu
http://www.physics.ucla.edu/faculty/ladder/peccei.html

Solid-State Physics, Editor

Peter Wölfle

Institut für Theorie der Kondensierten Materie
Universität Karlsruhe
Postfach 69 80
D-76128 Karlsruhe, Germany
Phone: +49 (7 21) 6 08 35 90
Fax: +49 (7 21) 69 81 50
Email: woelfle@tkm.physik.uni-karlsruhe.de
http://www-tkm.physik.uni-karlsruhe.de

Complex Systems, Editor

Frank Steiner

Abteilung Theoretische Physik
Universität Ulm
Albert-Einstein-Allee 11
D-89069 Ulm, Germany
Phone: +49 (7 31) 5 02 29 10
Fax: +49 (7 31) 5 02 29 24
Email: steiner@physik.uni-ulm.de
http://www.physik.uni-ulm.de/theo/theophys.html

Fundamental Astrophysics, Editor

Joachim Trümper

Max-Planck-Institut für Extraterrestrische Physik
Postfach 16 03
D-85740 Garching, Germany
Phone: +49 (89) 32 99 35 59
Fax: +49 (89) 32 99 35 69
Email: jtrumper@mpe-garching.mpg.de
http://www.mpe-garching.mpg.de/index.html

Wilhelm Kulisch

Deposition of Diamond-Like Superhard Materials

With 60 Figures

Springer

Dr. Wilhelm A.M. Kulisch
University of Kassel
Institute of Technical Physics
Heinrich-Plett-Strasse 40
D-34109 Kassel, Germany
Email: kulisch@physik.uni-kassel.de

Physics and Astronomy Classification Scheme (PACS):
62.20.-x, 68.60.-p, 81.05.Se, 81.05.Zx, 81.15.Zx, 81.15.-z

ISSN 0081-3869
ISBN 3-540-65629-4 Springer-Verlag Berlin Heidelberg New York

Library of Congress Cataloging-in-Publication Data.
Kulisch, Wilhelm. Deposition of diamond-like superhard materials/Wilhelm Kulisch. p. cm. – (Springer tracts in modern physics, ISSN 0081-3869; v. 157). Includes bibliographical references and index. ISBN 3-540-65629-4 (alk. paper). 1. Diamonds, Artificial 2. Chemical vapor deposition. I. Title. II. Series: Springer tracts in modern physics; 157. QC1.S797 vol. 157 [TP873.5.D5] 539 s–dc21 [666'.88] 99–26573

This work is subject to copyright. All rights are reserved, whether the whole or part of the material is concerned, specifically the rights of translation, reprinting, reuse of illustrations, recitation, broadcasting, reproduction on microfilm or in any other way, and storage in data banks. Duplication of this publication or parts thereof is permitted only under the provisions of the German Copyright Law of September 9, 1965, in its current version, and permission for use must always be obtained from Springer-Verlag. Violations are liable for prosecution under the German Copyright Law.

© Springer-Verlag Berlin Heidelberg 1999
Printed in Germany

The use of general descriptive names, registered names, trademarks, etc. in this publication does not imply, even in the absence of a specific statement, that such names are exempt from the relevant protective laws and regulations and therefore free for general use.

Typesetting: Camera-ready copy by the author using a Springer LATEX macro package
Cover design: *design & production* GmbH, Heidelberg
Computer-to-plate and printing: Mercedesdruck, Berlin
Binding: Universitätsdruckerei H. Stürtz AG, Würzburg

SPIN: 10697037   56/3144/tr - 5 4 3 2 1 0 – Printed on acid-free paper

Preface

Modern key technologies such as microelectronics and micro system technology, but also automobile, aircraft and space technology, and in addition even the fabrication of every-day articles (especially for recreational purposes) place an ever increasing demand on the materials involved. This is true not only for the work pieces and components produced (such as active or passive devices in microelectronics, or parts of cars and planes) but also for the tools used to fabricate these elements, and for components of the analysis tools used to characterize them. An example for the importance of material selection can be found in the automobile industry where harder cutting and spaning tools result not only in prolonged life of the (often very expensive) tools but also in smaller tolerances of the dimensions of motor components and thus to less expensive and less polluting motors. In the field of analytics, the recently developed scanning probe microscopies may be used as an example of tools which allow material characterization with regard to a variety of physical and chemical properties but which rely to a large extent on the material properties of the sensors used [1].

Thereby, it is sufficient for many (but not for all) applications to optimize not the bulk properties of a material but just its surface properties. In many cases this is more economical and often also leads to even better results. Such a surface optimization can be achieved either by chemical or physical surface treatments (plasma oxidation, plasma nitridation, ion implantation), or by deposition of a thin film with the desired properties.

For these reasons materials sciences in general, and especially thin film technology, experienced a rapid development in the two preceding decades. In this context, even actual physical topics gain more and more importance for materials sciences; the deposition of diamond films and the synthesis of high-temperature supraconductors, fullerenes, and nanotubes may serve as examples. In its initial phase, the development of thin film technology was closely related to that of semiconductor electronics; meanwhile, however, the fields of application have been extended widely and now cover areas such as optics, tribology, information storage, and many others.

A very important material property currently the center of intensive multifaceted research is hardness, which is the resistance against the intrusion of another body. Possible areas of applications again include work pieces and components (e.g. parts for aircrafts and boats), tools (e.g. drills, cutting tools), and components of analysis tools (e.g. resistant tips for AFM measurements) alike. It is further of special importance that the material property hardness correlates in many cases with other extreme mechanical and non-mechanical properties.

There are two general concepts in the design of superhard materials. The conventional way is to prevent the movement of dislocations and the propagation of cracks within the material. This approach, which has been known for several thousand years, today leads in the form of composite materials or – by thin film technologies – of multilayer and multiphase systems to new, successful solutions.

The second way consists of the synthesis of ideal materials (almost) free of dislocations and cracks. Therefore, besides the realization of such materials the question of importance is which compounds possess, due to their bond structure, an 'intrinsic' hardness that is as high as possible. In this context the prediction of material properties by means of ab initio calculations plays an ever increasing role [2]. The discussion in this book will show that superhard materials are to be found in the B/C/N system, and that they possess properties similar to those of diamond even if they do not exceed them. In addition to diamond, we also have tetrahedral amorphous carbon (ta-C) and cubic boron nitride (c-BN) in the class of diamond-like superhard materials. Currently, however, intensive work is going on in the race to synthesize diamond-like superhard carbon nitride modifications and – although this is still in its infancy – to realize ternary $B_xC_yN_z$ compounds.

This book summarizes the present state of the art of the deposition of diamond-like superhard materials in form of thin films. The experimental and theoretical work performed by the Thin Film Technology Group of the Institute for Technical Physics at the University of Kassel is used as a basis. In particular I discuss whether there is a common route to the deposition of such thin diamond-like superhard films which can be easily transferred to further compounds from this class, thus allowing the design of a family of superhard materials with many additional extreme properties.

Much of the content of this book comes from my German habilitation thesis, which was submitted to the Department of Physics at the University of Kassel in January 1998. Although research in the area of diamond-like superhard materials has by no means ceased, and a great number of interesting papers were published in 1998, the main conclusions of this book are not affected by this new work. I have included up-to-date references for readers who would like to investigate this field further.

Acknowledgement. Such a book cannot be written without the help and assistance of many people. Far more important is the fact that the underlaying experimental and theoretical research relies on the individual work of my coworkers and the help of and discussions with many friends and partners all over the world.

Preface   VII

First of all, I would like to thank from my heart my coworkers at the Institute of Technical Physics at the University of Kassel who performed the work I summarize in this book: L. Ackermann, D. Albert, A. Baranyai, R. Beckmann, R. Freudenstein, A. Klett, M. Kuhr, S. Reinke, and B. Sobisch. Likewise I thank Prof. R. Kassing (head of the Institute of Technical Physics) for a fruitful and trustful cooperation which has now lasted for more than 15 years, for countless intensive discussions, and, most important, for the freedom to perform this research according to my own ideas. I further thank A. Bock, K. Ewert, R. Linnemann, K. Masseli, Ch. Mihalcea, S. Münster, E. Oesterschulze, M. Stopka, S. Werner, W. Scholz, and, especially, G. Leber and A. Malkomes, from the Institute of Technical Physics at the University of Kassel, for their cooperation and assistance. The University of Kassel supported my research financially in the framework of a variety of projects.

This work was also financially supported by the German Research Society (DFG) under the 'Schwerpunktprogramm Synthese Superharter Materialien' as part of the trinational 'D-A-CH' German, Austrian, and Swiss cooperation on the 'Synthesis of Superhard Materials'. Almost even more important than the financial support from the DFG was the cooperation with many partners from the program. I therefore would really like to thank the following partners from the 'D-A-CH community' for cooperation, discussions, exchange of samples, performance of characterizations, and other assistance: C. Benndorf, P. Joeris, I. Schmidt, S. Hadenfeldt (University of Hamburg) E. Jansen, E. Obermeier (Technical University Berlin), R. Kuschnereit, P. Hess (University of Heidelberg), M. Schreck, K.-H. Thuerer, B. Stritzker (University of Augsburg), E. Blank, Th. Lang (EPFL Lausanne) G. Lippold, W. Grill (University of Leipzig), H. Ehrhardt, S. Ulrich, W. Dworschak F.R. Weber, H. Oechsner (University of Kaiserslautern) H. Hofsäss, C. Ronning (University of Konstanz), G. Dollinger, A. Bergmaier (Technical University Munich), H. Sachdev (University of Saarbrücken), H. Noeth (University of Munich), B. Lux, S. Bohr, R. Haubner, W. Kalss (Technical University Vienna) Th. Frauenheim, J. Widany, F. Weich (Technical University Chemnitz) J. von der Gönna, G. Nover, G. Will (University of Bonn), H. Sternschulte, P. Ziemann (University of Ulm).

Further financial support was provided by the European Commission. I would therefore like to thank N.E.W. Hartley (European Commission) and M.P. Delplancke-Ogletree, R. Winand, L. Segers, F. Kiel, and M. Cotarelo (Universite Libre de Bruxelles) for the cooperation in the framework of this project.

Finally, I have to thank the following partners for discussions, hints, exchange of samples, performance of analyses, and technological assistance: J. Wilson, M. Jubber, M. Liehr, D. Milne (Hariot–Watt University Edinburgh), Ch. Wild, P. Koidl (Fraunhofer Institute for Applied Solid State Physics, Freiburg), K. Bewilogua, S. Jäger, X. Jiang (Fraunhofer Institute for Film and Surface Techniques, Braunschweig), J. Hahn, F. Richter (Technical University Chemnitz), J. Robertson (University of Cambridge), P.B. Mirkarimi, D.L. Medlin, K. McCarty (Sandia, Livermore), D. Bouchier (University Paris–Sud), N. Tanabe (RIMES LTd, Tokyo), W.A. Yarbrough (Pennsylvania State University) J.C. Angus, W.A.L. Lambrecht (Case Western University, Cleveland), E. Schultheiss, C. Rau (IMO Wetzlar).

Kassel, April 1999   *Wilhelm Kulisch*

Contents

1. **The Material Property Hardness** 1
   1.1 Definition of Hardness 1
   1.2 Hardness Measurements 2
   1.3 Hard Materials 5
       1.3.1 Ideal Materials 5
       1.3.2 Real Materials 8
   1.4 Correlations with Other Material Properties 12
       1.4.1 Hardness versus Volumetric Lattice Energy 12
       1.4.2 Hardness and Bulk Modulus 13
       1.4.3 Model of Kisly 15
       1.4.4 Model of Cohen 16

2. **Diamond-Like Materials** 21
   2.1 Materials 21
   2.2 The B/C/N System 21
       2.2.1 Materials in the B/C/N System 22
       2.2.2 Carbon and Boron Nitride 23
   2.3 Properties of Diamond-Like Materials 27
       2.3.1 Thermal and Vibrational Properties 29
       2.3.2 Optical and Electronic Properties 32
   2.4 State of the Art 35
       2.4.1 Diamond 35
       2.4.2 Tetrahedral Amorphous Carbon 36
       2.4.3 Cubic Boron Nitride 38
       2.4.4 β-C_3N_4 40
   2.5 Organization of the Remainder of this Book 41

3. **Growth Mechanisms** 43
   3.1 On the Deposition of Thin Films 43
   3.2 Growth of Diamond Films 44
       3.2.1 Methods for Low-Pressure Diamond Synthesis 44
       3.2.2 Gas Phase Processes 47
       3.2.3 Transport Processes 59
       3.2.4 Surface Processes 63

		3.2.5	Macroscopic Diamond Growth	74
		3.2.6	Conclusions	77
	3.3	Growth of c-BN Films		79
		3.3.1	Methods and Parameters	79
		3.3.2	Gas Phase and Transport Processes	87
		3.3.3	Ion-Induced Mechanisms	88
		3.3.4	Modeling of c-BN Growth	89
		3.3.5	The Sputter Model	93
		3.3.6	Conclusions	101
	3.4	Growth of ta-C Films		105
		3.4.1	Experimental Observations	106
		3.4.2	Growth Mechanisms	108
		3.4.3	Conclusions	111
	3.5	Growth of CN Films		112
		3.5.1	Experimental Observations	112
		3.5.2	Discussion	117
	3.6	Conclusions		120
4.	**Nucleation Mechanisms**			**125**
	4.1	On the Theory of Nucleation		125
		4.1.1	Macroscopic Thermodynamic Description	125
		4.1.2	Statistical Thermodynamic Description	128
		4.1.3	Atomistic Description	129
		4.1.4	Influence of Surface Defects	131
		4.1.5	(Hetero)epitaxy	132
		4.1.6	Nucleation of Superhard Materials	133
	4.2	Nucleation of Diamond		134
		4.2.1	General Considerations	134
		4.2.2	Nucleation on Untreated Substrates	136
		4.2.3	Nucleation on Highly-Oriented Graphite	138
		4.2.4	Abrasive Treatment	140
	4.3	Bias-Enhanced Nucleation of Diamond		141
		4.3.1	General Considerations	141
		4.3.2	Experimental Observations	144
		4.3.3	Discussion and Modeling	155
	4.4	Nucleation of c-BN		159
		4.4.1	Experimental Observations	160
		4.4.2	Mechanisms of Nucleation	163
	4.5	Nucleation of ta–C and β-C_3N_4		167
	4.6	Conclusions		168
References				**173**
Index				**187**

1. The Material Property Hardness

1.1 Definition of Hardness

Hardness is defined as the resistance of a body against the intrusion of another (harder) body. It is a complex and complicated property [3]. The measured hardness depends, on the one hand, on the elastic and plastic properties of the material to be investigated but also, on the other hand, on the measuring technique applied and on the shape and nature of the indenter. Despite many attempts to develop a strong physical definition of hardness and, thus, also an absolute hardness scale, the theory of hardness has remained up to now semiempirical [4, 5, 6]. As a consequence, up to now it has also proven impossible to correlate the hardness of a material strongly to other material properties, although for individual classes of materials semiempirical relationships have been developed between hardness on the one hand and properties such as the nature of the bonding, the bond strength, the bond length, etc. on the other hand, which are discussed in detail in Sect. 1.4.

For the following discussion in this chapter two other material properties are also of importance, which are – as will be shown – closely related to hardness and which are often used synonymously with 'hardness', but which nevertheless have other physical meanings: the (tensile) strength σ_T, and the compressibility β and its reciprocal, the bulk modulus K.

The tensile strength σ_T is the maximum stress occuring during a tensile test, related to the original cross-sectional area of the body under test. Like hardness, it depends in a complex manner on the elastic and plastic properties of the material [7, 8]; nevertheless, as shown in Sect. 1.3.1, one can at least derive a theoretical value of the tensile strength of ideal materials.

In contrast to hardness and tensile strength, unequivocal physical definitions exist for the compressibility and bulk modulus; β is the isothermal relative volume change of a body produced by the application of a hydrostatic pressure p:

$$\beta = -\frac{1}{V}\left(\frac{\partial V}{\partial p}\right)_T. \tag{1.1}$$

For the bulk modulus K it follows therefore (where U is the internal energy) that

$$K = \frac{1}{\beta} = -V\frac{dp}{dV} = V\frac{d^2U}{dV^2} \ . \tag{1.2}$$

At this point it should be noted that the tensile strength and compressibility of a material represent bulk properties; hardness, on the other hand, is a surface property. The hardness of an object can therefore – in contrast to its strength and compressibility – be improved by an appropriate surface treatment, in particular by the deposition of a thin film of a hard or superhard material[1].

1.2 Hardness Measurements

In the following, the most important methods of measuring the hardness of a material will be discussed briefly; in this context, the lack of a physical definition of the quantity 'hardness' will again become apparent.

The measured hardness depends not only on the elastic and plastic properties of the material to be investigated, but also on the method applied and the parameters of the measurement. As a consequence, different hardness scales exist for the different techniques, which cannot be compared straightforwardly.

Table 1.1. The materials of the Mohs hardness scale. For comparison, the values of the Vickers hardness are given also [9]. 1000 kp/mm^2 = 9.81 GPa

H_M	Material	H_V (kp/mm^2)	H_M	Material	H_V (kp/mm^2)
1	Talc	2.4	6	Orthoclase	800
2	Gypsum	36	7	Quartz	1120
3	Calcite	110	8	Topaz	1430
4	Fluorite	190	9	Corundum	2000
5	Apatite	540	10	Diamond	(10 000)

The oldest of these scales was introduced by Mohs: A solid is harder than another if it can be used to scratch this other material. It has to be pointed out, however, that in a scratch test the material to be investigated not only is impressed but can also be partially removed. In this respect this test does not comply entirely with the above definition of hardness.

The scale introduced by Mohs relies on ten selected materials to which hardness values between 1 and 10 are attributed (Table 1.1). It has to be

---

[1] To this end, relatively thick films in the range of some microns are required, in contrast to the modification of other surface properties for which much thinner films are sufficient.

mentioned, however, that this scale has been chosen somewhat unfortunately; in comparison e.g. with the Vickers hardness (see below), the materials with Mohs hardnesses between 1 and 9 cover the region up to $H_V \approx 2000$, whereas all harder materials are classified between $H_M = 9$ and $H_M = 10$; these materials, however, can possess Vickers hardnesses up to 10 000 (type IIa diamond). The original scale of Mohs has therefore been refined and extended in the upper region (e.g. [4, 5]); in this revised form it is, even today, still used for the characterization of materials as well as for theoretical considerations [4, 5].

Table 1.2. Compilation of some important indentation methods for the measurement of hardness [10, 11, 12]. F denotes the applied load

Method	Indenter	Material	Definition and notes
Brinell	Sphere	Steel, tungsten carbide	$H_B = \dfrac{2F}{\pi D(D - \sqrt{D^2 - d^2})}$ D = sphere diameter, d = diameter of impression. The measured hardness depends on the applied load.
Rockwell	Sphere, cone	Steel, diamond	The Rockwell hardness is the difference between the impression depths after indenting with low and high loads. Depending on the form and material of the indenter and on the starting and final loads there are alltogether 18 Rockwell scales covering different ranges of hardness [10].
Vickers	Quadratic pyramid	Diamond	$H_V = \dfrac{1.854F}{d^2}$ d = diameter of impression. Owing to the form of the indenter, the hardness is independent of the applied load.
Knoop	Rhombohedral pyramid; ratio of diagonals 7:1	Diamond	$H_K = \dfrac{14.2F}{l^2}$ l = length of the long diagonal of the impression. Owing to the long diagonal this method has advantages with respect to the Vickers in the case of very hard coatings and low test loads [13].
Berkovitch	Triangular pyramid	Diamond	Values are mostly 10–20 % higher than Knoop hardness values [13].

All other methods are directly related to the above definition of hardness, i.e. the resistance against intrusion by another, harder material: an indenter is pressed with a defined force into the material to be investigated, and the area of the resulting impression is measured[2]. Table 1.2 compiles the most important techniques; they are distinguished basically by the form and material of the indenter. For the measurement of bulk materials, the Brinell and Rockwell techniques are usually applied [10, 12]; the measurement of the hardness of thin films, on the other hand, is in most cases carried out by means of the Vickers and Knoop methods, because of the lower forces applied [11]. However, it must be pointed out that the results obtained by different techniques are by no means consistent; in addition, even for a given technique the measured hardness often depends on the applied force and in some cases also on the impression time[3].

In the literature, there exist a variety of empirical relationships between the Mohs hardness H_M and the Vicker hardness H_V of a material, e.g.

$$H_V = 86.3 - 90.9 H_M + 34.6(H_M)^2 \quad \text{or} \quad H_V = 5.25(H_M)^{2.73}, \qquad (1.3)$$

of which the latter fits the existing data better [5]. Nevertheless, a strong correlation is observed only for the 'index materials' listed in Table 1.1; for the bulk of the data a considerable scatter exists, which is mostly due to the fact that the measurements of Vickers hardness which form the basis of these data collections were by no means carried out with the same parameters [5]. This underlines again the problems concerning an exact definition of hardness, and, as a consequence, of an absolute hardness scale.

Finally, a special case should be addressed very briefly which is of the greatest importance for this book: the body to be investigated is not a homogeneous material but, rather, a thin-film/substrate system. In this case, the results of hardness measurements always depend on the elastic and plastic properties not only of the coating but also of the substrate, and, in addition, on the nature of the interface between them; thus, a situation which is complex anyway is further complicated. As a rule of thumb it is generally agreed that in order to investigate the hardness of a coating alone, the depth of the impression of the indenter should be less than one tenth of the thickness of the coating[4]. Even in this case, however, an influence of the substrate and interface on the result of a hardness measurement cannot generally be excluded. As a consequence, the evaluation and interpretation of hardness measurements of film/substrate systems always require special care; many of the data published in the literature concerning the hardness of thin coatings

---

[2] In fact, the projection of the impression area on the surface is measured; however, the real impression area which is taken into account by the formulas given in Table 1.2, is the decisive quantity.

[3] 5400 H_K 0.5/20 means a hardness of 5400 kp/mm^2, measured by the Knoop method with an applied force of 5 N (0.5 kp) and an impression time of 20 s. 1000 kp/mm^2 is approximately 10 GPa.

[4] For applications, on the other hand, the hardness of the whole system is the relevant property.

are therefore of limited value (see also the data collection of Friedrich et al. [14]).

For these reasons, the techniques of microhardness and ultramicrohardness measurements are applied more and more to determine the hardness of film/substrate systems; these terms do not refer to the physical dimensions of the materials to be investigated but rather to the test forces applied, which are between 0.02 and 2 N, and 1×10^{-5} and 2×10^{-2} N, for microhardness and ultramicrohardness, respectively.

In addition, in recent years the method of nanoindentation [15] has been increasingly applied to investigate the hardness of substrate/layer systems. In this case also a hard, well-defined indenter is pressed into the sample with low forces. However, not only the area of the resulting impression is measured; additionally, the load/displacement curve is recorded during loading as well as during unloading. From this curve, not only the hardness but also further mechanical properties such as the Young's modulus can, in principle, be determined. Nevertheless, even in the case of nanoindentation the results obtained depend in a complex manner on the properties of the substrate, layer, and interface [15], necessitating again a careful interpretation of the measurements.

1.3 Hard Materials

The irreversible intrusion of one material into another, as used to measure hardness, relies on the plastic deformation and/or brittle failure of the material under the influence of mechanical stresses. This implies a close relation between the surface property, hardness, and the bulk property, tensile strength. The plastic deformation, in the case of ductile materials, is essentially the result of the movement of dislocations; brittle failure is caused by the propagation of cracks. From this, two basically different concepts for the fabrication of hard (strong) materials follow immediately:

- The avoidance (or reduction) of defects, especially of dislocations and cracks, i.e. the fabrication of single crystals with a perfection as high as possible
- Measures to prevent the movement of dislocations and the propagation of cracks.

1.3.1 Ideal Materials

Every solid possesses an ideal (theoretical) strength which is determined by the nature of its chemical bonds [8]. This requires, however, an ideal solid without surfaces, cracks or dislocations. An infinite, perfect crystalline polymorph should thus represent the strongest form of a solid and therefore also the upper limit of the tensile strength (and hardness) which could be reached by the corresponding real material [8].

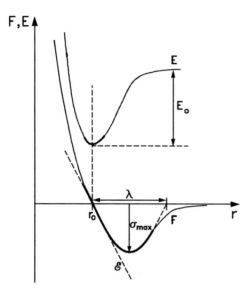

Fig. 1.1. Potential energy and interatomic force as a function of the distance between two atoms

The theoretical (ideal) tensile strength of a material can be estimated roughly by the following simple approach [7, 8]. The starting point is the interatomic force/distance curve shown in Fig. 1.1. In the region $r > r_0$, this curve can be approximated by a half sine wave:

$$\sigma = \sigma_{\max} \sin\left(\frac{\pi}{\lambda}(r - r_0)\right) . \tag{1.4}$$

The amplitude of this half sine wave $\sigma_{\max}$ is equivalent to the force which is required to separate two atoms, or in this case to the theoretical stress which is necessary to separate two neighboring atomic planes. It can be determined by making use of the definition of Young's modulus as the slope of the curve in the vicinity of r_0 (Fig. 1.1):

$$\mathcal{E} = r_0 \left(\frac{d\sigma}{dr}\right)_{r=r_0} = \sigma_{\max} \frac{\pi r_0}{\lambda} \longrightarrow \sigma_{\max} = \frac{\mathcal{E}\,\lambda}{\pi\,r_0} . \tag{1.5}$$

The 'range of the interatomic force' λ, still unknown in this equation, can be estimated by assuming that the work required to separate two atomic planes equals the surface energy 2γ of the newly generated surfaces. This work, however, is just the area under the half sine wave of the force/distance curve:

$$\int_{r_0}^{r_0+\lambda} \sigma\,dr = 2\gamma \longrightarrow \lambda = \frac{\pi\gamma}{\sigma_{\max}} . \tag{1.6}$$

From this, it follows that the theoretical fracture stress is

$$\sigma_{max} = \sqrt{\frac{\mathcal{E}\gamma}{r_0}}. \tag{1.7}$$

It can be seen that the higher the Young's modulus and surface energy and the lower the interatomic spacing of a material are, the higher σ_{max} is. These properties, however, are not independent of each other but strongly correlated (see Sects. 1.4 and 2.3).

For a tensile test with an ideal material, however, one has to take into account not only the separation of atomic planes but also the sliding of atoms on a slip plane under the influence of a shear stress [8]. In analogy to the above discussion it can be shown that the theoretical critical shear stress τ_{max} for such a process is

$$\tau_{max} = \frac{Ga}{2\pi h}, \tag{1.8}$$

where a is the spacing of the atoms within the slip plane and h is the spacing of these planes. Thus, τ_{max} is determined by the shear modulus G and the packing of the atoms. Since during a sliding process no new surfaces are generated and bonds are not only broken but also newly formed, τ_{max} is in general smaller than σ_{max} [8].

Table 1.3. Theoretical tensile strength σ_{max} and theoretical critical shear stress τ_{max} for the principal types of materials. The term 'high' should be taken to imply only that there are no limitations in principle

Type of bond	σ_{max}	τ_{max}
Ionic	Low owing to long-range repulsive forces	High
Metallic	High	Low owing to the nondirectional nature of bonds
Covalent	High	High

The highest values of σ_{max} are found for covalent and metallic solids (Table 1.3); in ionic crystals, on the other hand, not only attractive but also repulsive long-range interatomic forces exist, which tend to neutralize each other in part; as a consequence, σ_{max} is usually lower [8].

The ideal critical shear stress, on the other hand, is high for those materials that possess highly directional bonds. For this reason, metals usually have low values of τ_{max} (Table 1.3). An ideal strong material should possess high values of both σ_{max} and τ_{max}; according to Table 1.3 these are the covalent materials[5].

---

[5] This does not mean that all covalent materials automatically possess high strengths; rather, the term 'high' in Table 1.3 has to be interpreted in the sense

Measurements with specially prepared, almost defect-free solids (e.g. whiskers) show that (1.7) predicts the order of magnitude of the strength quite correctly but overestimates it by a factor of the order of 2 [8]. For most of the materials used in practical applications, however, the strength is lower than predicted by (1.7) by at least two orders of magnitude. As already pointed out, this is due to the existence of defects (dislocations and cracks) within these real materials. In the next section, therefore, measures to improve the strength (or the hardness) of materials despite the presence of such defects are discussed. The question of which materials possess an extremely high theoretical strength (hardness) according to (1.7) will be considered further in Sect. 1.4.

1.3.2 Real Materials

The real hardness or strength of a solid is – as we have seen – usually lower by orders of magnitude than in the ideal case described above. The preparation of defect-free, almost perfect materials the strength of which approximates the theoretical value (1.7) is in most cases too expensive or only possible for dimensions without technological relevance (e.g. whiskers)[6]. Therefore, for many (thousands of) years a wide variety of methods for the fabrication of hard materials has been developed which rely on the second of the concepts discussed above, without which the world in its present form would be impossible. In the following, some of these standard methods to prevent the movement of dislocations or the propagation of cracks are listed (without claiming completeness)[7]:

- Ductile materials become stronger when they are subjected to plastic deformation. This is caused by the multiplication of dislocations, for example by the Frank–Read mechanism (see below), and the repulsive interaction of dislocations with one another, by means of which dislocations

---

that there are limitations in principle. According to Kelly and MacMillan [8], in addition to highly directional bonds a short bond length and a high coordination number are required for high values of σ_{max} and τ_{max}, i.e. a dense three-dimensional network of strong bonds. The elements which are able to fulfill these conditions are Be, B, C, N, O, Al and Si; the strongest materials contain at least one of these elements and are often completely composed of them (see Sect. 1.4).

[6] At this point it should be pointed out again that the hardness of an object can be improved drastically by the deposition of a thin coating. In many cases, this is the most convenient solution. The tensile strength of a material, on the other hand, is not affected by such a coating.

[7] At this point it should be pointed out again that hardness and strength describe different physical properties although the terms are often used synonymously. Hardness is the resistance against intrusion by another material, the tensile strength is the resistance against mechanical (tensile) stresses. Nevertheless, movement of dislocations and propagation of cracks play most important roles during hardness measurements as well as during tensile tests.

mutually prevent their movement. The strength of a material can thus be improved by work hardening, i.e. plastic deformation.
- Grain boundaries act as physical barriers preventing the movement of dislocations, which therefore accumulate at grain boundaries. Accordingly the yield stress σ_y of a material which marks the onset of plastic deformation depends on the concentration of grain boundaries or the grain diameter d_g. This is described by the Hall–Petch equation:

$$\sigma_y = \sigma_i + k_y d_g^{-1/2} \ . \tag{1.9}$$

Here, $\sigma_i(T)$ and $k_y(T)$ are material constants describing the intrinsic resistance against the movement of dislocations and the efficiency of dislocation sources, respectively. The Hall–Petch equation is, however, valid only for grain diameters down to approximately 5 μm; below this value, a saturation effect is observed [16].
- Foreign atoms in solution (substitutional as well as interstitial) also impede the movement of dislocations and lead therefore to an increase of the yield stress. Like dislocations, they cause local elastic strains; often they accumulate at grain boundaries in such a way that these strains cancel each other out leading to a minimization of the total energy [10]; it is therefore difficult to move the dislocations away from the solute atoms. The best-known example of hardening by interstitial atoms in solution is that of carbon in b.c.c. iron.
- Finally, the presence of a second phase in the form of a dispersion of small particles also hinders the movement of dislocations in many cases and leads therefore to a considerable increase of the strength or the hardness. The volume fraction of the second phase and the particle size, which can be combined into the interparticle distance d_p, are decisive in this case .

These classical concepts for improving the mechanical properties of materials, i.e. work hardening and the incorporation of grain boundaries, solute atoms and particles, have become more and more refined; it can be taken for sure that there will be further improvements in the future. However, strength and hardness are not the only material properties that are important in a given application. In many cases a material must fulfill a long list of requirements. In addition, economic aspects play a most important role. As a consequence, the fabrication of a homogeneous hard material possessing simultaneously a variety of other special properties is, in many cases, much too expensive. These demands lead, among others, to the development of composites, especially the class of fiber-reinforced composites [17], which in general combine a high strength (hardness) and a high stiffness with a low specific weight[8].

---

[8] The hardness of a material can be improved by the deposition of thin films as has already been mentioned. Nevertheless, in general such films must also fulfill a variety of requirements. However, by a suitable choice of the deposition parame-

In the context of the development and improvement of hard materials based on the concept of hindering the movement of dislocations and the propagation of cracks, the following examples of current research activity may be somewhat arbitrarily – mentioned at this point.

Multilayer Coatings and Superlattices. Composition-modulated thin films (multilayer systems and superlattices) in many cases possess improved mechanical properties (e.g. hardness and Young's modulus) as compared to alloys of the same average composition, if the thickness λ of the individual layers is in the nanometer range [16, 19]. For example, Mirkarimi et al. [19] showed that the hardness of single-crystal, lattice-matched $TiN/(V_{0.6}Nb_{0.4})N$ layers with $\lambda \approx 8$ nm exceeds the hardnesses of the individual components (TiN and $(V_{0.6}Nb_{0.4})N$) by a factor of 2–2.5, reaching values up to $H_V \approx 4000$. For larger and smaller values of λ the hardness decreases. Similar results have also been obtained for polycrystalline metal/metal nitride multilayer systems (metal = Ti, Hf, W) [20]; their hardnesses can greatly exceed those of the relatively soft metals and exceed those of the hard metal nitrides by a factor of 1.5–2. In this case also, the hardness depends on the thickness of the individual layers: for large values of λ a relationship $H \propto \lambda^{1/2}$ in analogy to the Hall–Petch equation (1.9) has been found. The maximum hardness values are observed for $\lambda \approx 4$–20 nm; for the hafnium system, values up to $H_V \approx 5000$ can be reached.

The reasons for these excellent mechanical properties of composition-modulated coatings and, especially, the dependence on the thickness λ of the individual layers are not understood in detail [16, 19]. There is considerable agreement in the literature that the decrease of hardness for very small λ results from the increasing influence of the definition of the interfaces between the individual layers (interdiffusion). At higher λ, prevention of the movement of dislocations in analogy to the Hall–Petch mechanism, elastic strain at the interfaces and reflection of dislocations at the interfaces have, among others, been postulated as possible reasons for the increase of hardness with decreasing thickness.

Composites with Nanocrystals. According to Veprek et al. [16], composite materials consisting of nanocrystals in an amorphous matrix should possess extremely high hardnesses and extremely high Young's moduli; the limitation of strength at grain diameters of about 5 μm for polycrystalline one-phase materials, given by the range of validity of the Hall–Petch relation, can thus be exceeded. Nevertheless, two prerequisites have to be fulfilled: no

---

ters, the microstructure of a thin film can be influenced much more strongly than is possible in the fabrication of bulk materials. In addition, one has the possibility to deposit multilayers or multiphase systems in which each single component performs a different task. Criteria for the selection of materials for the deposition of hard coatings by 'classical methods' have been compiled by Holleck [18]. A great deal of further data on hardness and other properties of thin films can be found in [14].

intermixing between the two components takes place, and very sharp interfaces exist between them. Dislocations formed in the crystallites cannot move in the amorphous matrix; moreover, very small crystallites (nanocrystals) are even free of dislocations. The amorphous matrix, on the other hand, has a very high energy for the propagation of cracks, which is further hindered by the presence of the nanocrystallites. Finally, randomly oriented crystallites in an amorphous matrix should provide better coherence at the grain boundaries than polycrystalline materials.

As the first example of such nanocrystallite/amorphous-matrix composites, Vepřek et al. presented nc-TiN/a-Si_3N_4 (nc, nanocrystalline; a, amorphous) films with a hardness of about 50 GPa; the diameter of the nanocrystals has the most important influence on the hardness, which shows a maximum for diameters between 4 and 5 nm.

Composites with Nanotubes. The advantages of fiber-reinforced composites (e.g. carbon-fiber-reinforced epoxy resins) are – as already mentioned – their low density, simultaneously with high stiffness and high strength. These properties basically rely on two factors. The fracture strength of a brittle material depends on the dimensions of the largest internal crack within the material (see Sect. 1.4.3, (1.18)); the smaller the largest crack, the higher the strength. As a fiber cannot have transverse cracks larger than its diameter, fibers break at much higher stresses than, for example, glass. In addition to this intrinsic strength of the fibers, cracks present within the matrix are deflected at the relatively weak fiber/matrix interface. As long as the stress is directed parallel to the fibers, these cracks are harmless; the energy is absorbed by debonding of the fibers from the matrix and by friction effects during their pullout [10, 17].

According to Calvert [17], the discovery of nanotubes (Sect. 2.2.1) and of similar but slightly larger silicate tubes with a diameter which is smaller than that of conventional fibers by orders of magnitude opens new possibilities for the fabrication of extremely strong materials. As the fracture strength depends on the fiber diameter, composites reinforced by nanotubes should posses a higher strength than today's common materials.

These approaches – as interesting as they are – are nevertheless beyond the scope of this book, which deals with the hardness of ideal materials, thereby neglecting the problem of how to reach the required extreme crystalline perfection. Rather, at this point the question as to which materials possess, in the ideal case, a very high hardness (or a high tensile strength and low compressibility) has to be discussed; this in turn leads to the dependence of the hardness on other material properties such as the bond strength, nature of bonds and bond length.

1.4 Correlations with Other Material Properties

Although (or even because) the term hardness has no exact physical definition, there have been a variety of attempts published in the literature trying to correlate the hardness of a material with other properties, such as the interatomic distance, the atomic density, the valency of the elements involved, the nature of the bonds and the crystal structure[9]. In many cases people have tried to relate the hardness to the compressibility β or the bulk modulus K, which are – in contrast to hardness – defined unequivocally in a physical sense as described above. These studies are often motivated by an attempt to allow one to predict the hardness of a certain material or, vice versa, to propose extremely hard materials.

Nevertheless, all relationships proposed up to now in the literature, either between hardness and bulk modulus, or between hardness and/or bulk modulus and the other material properties listed above, possess some serious disadvantages, making it again evident that an unequivocal concept to describe hardness — and also the bulk modulus, for which after all a physical definition exists – has not yet been found:

- The proposed correlations are almost always purely empirical and rely in most cases on fits to selected sets of data.
- The correlations are almost always valid for certain classes of material only; for different classes, diverging relationships have often been found.

In the following, some of these studies are discussed briefly; of especial importance are the investigations of Cohen and coworkers [21, 22], as they initiated the current worldwide intensive search for superhard carbon nitride modifications.

1.4.1 Hardness versus Volumetric Lattice Energy

A first attempt to correlate the hardness with physical material properties was made by Plendl and Gielisse [4], which investigated the relationship between the Mohs hardness H_M and the lattice energy per unit volume U/V_m (volumetric lattice energy) for 65 semiconductors and insulators with a total of 16 different structures (V_m is the molar volume). They established a linear relation between U/V_m and H_M but nevertheless found it necessary to divide the investigated materials (in most cases binary compounds of the AB type) into two classes: the materials of type 1 are those in which at least one element has an atomic number less than 10 and both atomic numbers are lower than 18; type 2 includes all other materials.

With this classification, the following relationships between U/V_m and H_M have been found (in units of kcal/cm^3):

---

[9] A review of the parts of this literature not discussed below can be found in [5].

$$U/V_\mathrm{m} = 24(H_\mathrm{M} - 2.7) \qquad 4 < H_\mathrm{M} < 9, \text{ type 1}$$
$$U/V_\mathrm{m} = 48(H_\mathrm{M} - 3.3) \qquad 4 < H_\mathrm{M} < 9, \text{ type 2}$$
$$U/V_\mathrm{m} = 16(H_\mathrm{M} - 2.0) \qquad H_\mathrm{M} < 4, \text{ both types}. \qquad (1.10)$$

The difference of the slope for hardnesses less than and greater than $H_\mathrm{M} = 4$ seem to arise from the definition of the Mohs hardness scale[10].

Summarizing, it has to be pointed out that (1.10) is a purely empirical relationship; Plendl and Gielisse do not provide a phyical motivation; this also holds for the classification of the materials into two types. Finally, (1.10) is limited to semiconducting and insulating materials; metals and conductors are explicitly excluded.

1.4.2 Hardness and Bulk Modulus

In a similar manner, Goble and Scott [5] investigated on an empirical basis correlations between hardness, bulk modulus and volumetric lattice energy; in this context they point out especially that H and K depend on the same factors. In their investigation they considered the data for 81 solids which they divide into two classes, the alkali halides (from KBr to InSb) and the sulfides and oxides (e.g. AgBr and ZnO, but also TiN and SiC). A plot of the Mohs hardness versus the bulk modulus yields the following results:

- There is no unequivocal relationship between H_M and K valid for all classes of material.
- Nevertheless, there is a general but rather rough trend: the higher the bulk modulus, the higher the hardness.
- For certain classes of material (e.g. alkali halides and sulfides), on the other hand, special relationships seem indeed to exist between H and K.

A plot of H_M versus the volumetric lattice energy U/V_m shows that, in the case of the set of data used by Goble and Scott, the correlations (1.10) proposed by Plendl and Gielisse describe the relationship between these quantities correct only in terms of tendencies. A much better correlation exists, on the other hand, between the bulk modulus K and the volumetric lattice energy, corrected by the factor Zm/q, where q is the number of atoms per molecule, m is the number of components and Z is the maximum valency. Thus, the volumetric lattice energy is, via

---

[10] Plendl and Gielisse also give relations for materials with $H > 9$; in this range, however, they do not use the Mohs scale but the so-called Wooddell scale, which is related to the resistance against abrasion during lapping. As in the Mohs scale, corundum has here a value of 9; type II diamond, on the other hand, possesses a Wooddell hardness of 42.5 (instead of $H_\mathrm{M} = 10$). In this region the relation between H and U/V_m is again linear; the different hardness scales nevertheless cause again a change of slope. An explicit presentation of these relations is omitted for reasons of space.

$$K \propto \frac{Zm}{q}\frac{U}{V_m}, \tag{1.11}$$

a measure of the bulk modulus rather than – as suggested by Plendl and Gielisse – of the hardness.

Goble and Scott then postulate the following empirical relationships for the bulk modulus and hardness, again finding it necessary to divide the materials into two classes:

$$H_M = 15.38\frac{Z_e}{V_m^{2/3}}, \quad K = 6.65\frac{Z_c}{V_m} \quad \text{for alkali halides,} \tag{1.12}$$

$$H_M = 19.16\frac{Z_c^{2/3}}{V_m^{2/3}}, \quad K = 10.84\frac{Z_c^{3/4}}{V_m} \quad \text{for sulfides and oxides.} \tag{1.13}$$

Here, V_m is the molar volume, Z_c the valency of the cation and Z_e an effective valency, with $Z_e = 1 - (T_e - 12)/50.21$, where T_e is the total number of electrons of the compound. Goble and Scott conclude from these results that a relation indeed exists between hardness and bulk modulus; however, this correlation is rather indirect and depends on further terms, the physical significance of which is not clear. Finally, they postulate a physical definition of the hardness according to which hardness is the second derivative of the lattice energy with respect to the inter-atomic spacing, similar to the bulk modulus, which is the second derivative of the lattice energy with respect to the volume:

$$K = V\left(\frac{d^2U}{dV^2}\right)_{V_0} \quad \text{and} \quad H = \left(\frac{d^2U}{dr^2}\right)_{r_0}. \tag{1.14}$$

The dependence of the hardness on r instead of V reflects, according to Gobble and Scott, the fact that hardness – in contrast to the bulk modulus – is of a directional nature.

The studies of Plendl and Gielisse and Goble and Scott discussed in this section in some detail, as well as further similar investigations which are summarized in [5], once again illustrate the major points stressed throughout this chapter:

- There is no unequivocal physical definition of the quantity hardness.
- Between hardness, bulk modulus, volumetric lattice energy and other properties such as the molar volume (or the interatomic distance) correlations can indeed be found, which nevertheless are limited to special classes of material.
- A physical derivation of these correlations has proven impossible up to now.

1.4.3 Model of Kisly

According to Kisly [3], the work performed during a hardness measurement, i.e. the intrusion of an indenter into another material, includes the following four components:

1. **Elastic deformation.** The theoretical fracture stress which is necessary to separate two atomic planes has been derived in Sect. 1.3.1 for the ideal case, i.e. for materials without dislocations or cracks, as

$$\sigma_{max} = \sqrt{\frac{\mathcal{E}\gamma}{d}} \qquad (1.15)$$

   (where the bond length $d = r_0$).

2. **Plastic deformation.** Plastic deformation of a material is caused by the movement of dislocations under the influence of an applied stress. The critical shear stress necessary to multiply dislocations according to the Frank–Read mechanism is [23]

$$\tau_{FR} = \frac{2Gb}{d_0}, \qquad (1.16)$$

   where b is the Burgers vector of the dislocation and d_0 is the distance between neighboring slip planes.

3. **Formation and propagation of cracks.** According to an approach of Griffith [7], the energy balance during the propagation of a crack includes, on the one hand, the released strain energy $\sigma^2/2\mathcal{E}$ and, on the other hand, the surface energy of the newly created surface γ. For a crack of unit width and length L the total energy is thus

$$U_R = 2L\gamma - \frac{\pi L^2 \sigma^2}{4\mathcal{E}}. \qquad (1.17)$$

   U_R shows a maximum at the critical crack length L_c; for larger L the crack is unstable and minimizes the total energy by propagation. The critical stress σ_c causing the propagation of the crack for a given crack length is therefore given by

$$\frac{dU_R}{dL} = 2\gamma - \frac{\pi L \sigma_c^2}{2\mathcal{E}} = 0 \quad \longrightarrow \quad \sigma_c = \sqrt{\frac{\gamma \mathcal{E}}{\pi L}}. \qquad (1.18)$$

   This mechanism describes the response of brittle materials to the intrusion process; plastic deformation is observed exclusively in ductile materials. For brittle materials the hardness is therefore determined by the longest crack; in addition, it is higher the higher the Young's modulus and surface energy are.

4. **Creation of new surfaces.** For this term Kisly assumes that the hardness is proportional to the surface energy γ.

In summary, according to Kisly the following expression results for the hardness:

$$H = c_1 \sqrt{\frac{\mathcal{E}\gamma}{d}} + c_2 \frac{Gb}{d_0} + c_3 \sqrt{\frac{2\mathcal{E}\gamma}{L}} + c_4 \gamma \ . \tag{1.19}$$

Between the free-surface energy and the elastic constants of a material, on the one hand, and its bond energy E_0, its degree of covalency a_c and its bond length d, on the other hand, there exist the following relationships[11]:

$$\mathcal{E} \propto \frac{E_0 a_c}{d^2} \quad , \quad G \propto \frac{E_0 a_c}{d^3} \quad \text{and} \quad \gamma \propto \frac{E_0 a_c}{d^2} \ . \tag{1.20}$$

The degree of covalency of a bond is given by $a_c = 1 - a_i$, where a_i is the ionicity according to Pauling, which is determined by the electronegativities ϵ_A, ϵ_B of the elements A and B [25]:

$$a_i = 1 - \exp\left(-\frac{1}{4}(\epsilon_A - \epsilon_B)^2\right) \ . \tag{1.21}$$

Inserting these relations into (1.19) one obtains

$$H = C_1 \frac{E_0 a_c}{d^{5/2}} + C_2 \frac{E_0 a_c}{d^3} \cdot \frac{b}{d_0} + C_3 \frac{E_0 a_c}{d^2 \sqrt{L}} + C_4 \frac{E_0 a_c}{d^2} \ . \tag{1.22}$$

In this equation, all terms contain the quantity $E_0 a_c / d^2$; one can therefore summarize it as follows [26]:

$$H = \frac{E_0 a_c}{d^2} \left(C_1 \frac{1}{\sqrt{d}} + C_2 \frac{1}{d} \frac{b}{d_0} + C_3 \frac{1}{\sqrt{L}} + C_4 \right) \ . \tag{1.23}$$

Thus, the hardness of a material depends on its bond energy, the covalency of the bond and the bond length. In addition, parameters describing plastic deformation and brittle fracture are of importance (b, d_0 and L).

1.4.4 Model of Cohen

The discussion in Sect. 1.4.2 has shown that no general correlation valid for all materials exists between the hardness H and the bulk modulus K. It is found empirically, however, that for ideal, i.e. almost defect-free, isotropic systems, to which the discussion is restricted in the following, the hardness is proportional to K [6, 22, 27][12].

---

[11] At this point, Kisly refers to Harrison [24]. But again, he considers only certain classes of materials: covalent compounds and high-melting-point carbides, nitrides and oxides.

[12] Very recently, Teter suggested the shear modulus to be a better measure of hardness than the bulk modulus since hardness measurements always have a shear character [28].

For the bulk modulus of covalent solids, according to Cohen and Liu [2, 21, 22], the following semiempirical dependence on the bond length d, the mean coordination number $\langle N_c \rangle$ and the ionicity of the bond λ exists[13]:

$$K \ (\text{GPa}) = \frac{\langle N_c \rangle}{4} \frac{1971 - 220\lambda}{(d \ [\text{Å}])^{3.5}} \ . \tag{1.24}$$

This equation is also semiempirical at best; a physical derivation is not possible at present [29][14].

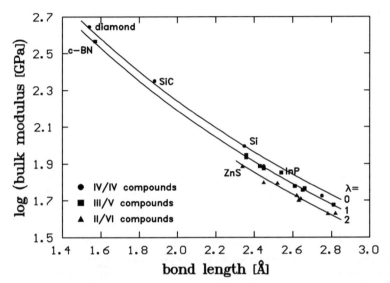

Fig. 1.2. Dependence of the bulk modulus K on the bond length d for some IV/IV, III/V and II/V compounds. The *solid lines* represent (1.24) for $\lambda = 0,1,2$. Data from [2]

In Fig. 1.2 the bulk moduli of some IV/IV, III/V and II/V compounds are plotted as a function of the bond length. The diagram shows that for these covalent compounds K is well described by equation (1.24). Simultaneously, the extreme influence of the bond length and the resulting eminent position of the materials diamond and c-BN become evident, whereas the influence of the ionicity on the bulk modulus is rather weak.

Materials with an extremely high bulk modulus and thus a high hardness are, according to (1.24), marked by a short bond length, a low ionicity and a

---

[13] λ has the values 0, 1 and 2 for IV/IV, III/V and II/VI compounds, respectively.
[14] In [21] Cohen presents arguments making the $d^{-3.5}$ dependence of K plausible. For reasons of space we refrain from a presentation of these considerations, which are based on the definition (1.2) of the bulk modulus and which are explicitly restricted to covalent materials.

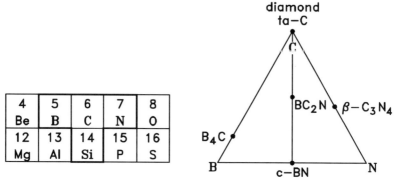

Fig. 1.3. *Left*: the elements in the vicinity of carbon in the periodic table. *Right*: superhard materials in the B/C/N system

high coordination number. Such materials can be found in the periodic table in the direct vicinity of carbon (Fig. 1.3). Superhard materials[15] should, according to the present state of knowledge, be restricted to the elements boron, carbon and nitrogen.

At the present time, only carbon (in crystalline form as diamond and in amorphous form as ta-C (see Sect. 2.4.2)) and cubic boron nitride (c-BN) can be regarded as superhard if the above definition is applied. Nevertheless, in 1989 Liu and Cohen [22] proposed on the basis of (1.24) that crystalline carbon nitride compounds with tetrahedrally bonded carbon should possess an extremely high bulk modulus. In analogy to the relatively hard, well-known and also technologically important β-Si_3N_4 , they suggested β-C_3N_4 as a possible candidate for such a superhard carbon nitride modification. According to ab initio calculations β-C_3N_4 should possess a bond length of only 0.147 nm and a bulk modulus of 427 GPa [22, 31, 32] (the value for diamond is 443 GPa, see Table 2.1).

Diamond is the hardest material known today; in addition, it is marked by a variety of other extreme properties, which are discussed in detail in Sect. 2.3. Superhard materials with short bond lengths, low ionicities and high coordination numbers, similar to diamond, will therefore be called 'diamond-like'[16] in the following. Like diamond, they possess a variety of other outstanding

---

[15] Definitions of the term 'superhard' are of course arbitrary. Usually, a Vickers hardness of $H_V = 4000$ is chosen as the lower limit [30]. Thus, materials such as B_4C ($H_V = 3000$–4000) and SiC ($H_V = 2500$–4000) are hard but not superhard [14, 30]. In the periodic table, these materials are also located in the direct vicinity of the element carbon. On the other hand, Kisly [3] classifies as superhard all materials with a Mohs hardness > 9; this condition is fulfilled for a large number of materials.

[16] The term 'diamond-like materials' has be chosen in accordance with [26]. In general, it applies to materials with physical properties similar to those of diamond, and should not be confused with the term 'diamond-like carbon' (DLC), which is usually used to label hard, but amorphous carbon layers [33] (see Sect. 2.4.2).

properties, in most cases close to but almost always somewhat short of those of diamond (Table 2.1 and Sect. 2.3). Therefore, these materials are of the utmost importance from both, a technological and a theoretical point of view.

This book deals, therefore, with the fabrication of diamond-like superhard materials in the form of thin films. In particular, the question is discussed of whether a common process or some common mechanisms for the deposition of this class of material can be found which can be transferred in future to other compounds in the B/C/N system which have not been investigated up to now or have been investigated only very little [34, 35, 36, 37], for example the ternary compounds $B_x C_y N_z$, which may possess new, interesting properties. Nevertheless, before dealing with this question, it is necessary at this point to discuss the diamond-like materials and their properties in some detail.

2. Diamond-Like Materials

2.1 Materials

Diamond-like superhard materials are to be found – as discussed in Sect. 1.4.4 – in the direct vicinity of carbon in the periodic table, i.e. within the system B/C/N. In this book, the following four materials are mainly considered:

1. diamond
2. tetrahedral amorphous carbon
3. cubic boron nitride
4. the hypothetical β-C_3N_4.

The book focuses on diamond and c-BN. Tetrahedral amorphous carbon is considered for comparison purposes since it is – like diamond – a pure carbon modification, while its fabrication nevertheless shows some parallels to the deposition of c-BN and differs distinctly from the low-pressure synthesis of diamond. It has to be taken into account, however, that ta-C is a not a crystalline but an amorphous material. Finally, β-C_3N_4 is a hypothetical compound which, according to calculations, should possess properties close to or even exceeding those of diamond [22]. All attempts to synthesize β-C_3N_4 have, however, failed up to now (or yield at best some extremely small, nanocrystalline quantities, see Sects. 2.4.4 and 3.5); it is therefore of interest to investigate whether the concepts of diamond and c-BN deposition can be transferred to β-C_3N_4, and what are the reasons for the failure of (almost) all synthesis attempts, in order to reach eventually some general conclusions on the synthesis of materials in the B/C/N system (e.g. ternary compounds).

2.2 The B/C/N System

Before discussing in some detail the properties of the diamond-like superhard materials and the state of the art concerning their deposition in the form of thin films, it seems to be worthwhile to mention briefly some peculiarities of the B/C/N system which are responsible for the current immense worldwide interest not only in the superhard phases diamond and c-BN but also in other materials from this system.

2.2.1 Materials in the B/C/N System

In contrast to the tetrahedral sp^3-bonded diamond, the sp^2-bonded graphite has no analog with the other elements of group IV (Si, Ge and Sn) [38, 39]. The reason can be found in the electronic structure of the core of the elements of the first row of the periodic table [38], which consists of s-electrons only and contains no p-electrons. Any p-electrons in the core have a repulsive effect on p-valence electrons, forcing them into bonds with neighboring atoms [40]. This means that sp^2 hybridization takes place only for elements of the first row. The ability to form double and triple bonds, however, is (besides the ability to form long chains and cyclic compounds) responsible for the immense variety of organic chemistry.

Since boron and nitrogen are also located in the first row of the periodic table, sp^2 hybridization is also feasible for boron nitride, and likewise for further (possible) binary and ternary compounds in the B/C/N system [38]. Owing to the ability to develop sp^2 bonds, $B_xC_yN_z$ compounds can therefore form graphite-like, layered modifications; in addition, they are possible candidates [38] for further materials which are at present intensively investigated theoretically as well as experimentally, such as fullerenes [41] and fullerites [42, 43, 44][1], as well as nanotubes (buckytubes) [45, 46, 47] and the entirely sp^2-bonded H6 carbon [48].

Thus, the current immense interest of material scientists in the B/C/N system is caused not only by the properties of the various tetrahedrally bonded materials but also by the facts listed above. This is true not only for application-oriented research and development but also for basic research and, especially, for theoretical work. The enormous development of ab initio calculations [2, 29] allows the determination of the lattice constants, bulk moduli, binding energies, phonon spectra, band gaps, etc. of a large class of materials. In addition, it enables for the first time the prediction of important properties of hypothetical materials in the B/C/N system which have not been realized up to now [22, 48]. It can be taken as certain that these methods will be further improved in the future.

The main topic of this book is the material property hardness. The sp^2 bond of carbon atoms is shorter than the sp^3 bond (Table 2.1). If it were be possible to compress a graphite layer without allowing the atoms to move in the c direction, the compressibility would be lower than that of diamond [38]. On the other hand, a three-dimensional, completely sp^2-bonded solid (H6 carbon [48]) would possess a coordination number of only three; as a

---

[1] The formation of BN fullerenes and fullerites is not possible since fullerenes necessarily contain – besides a variable number of six-membered rings – twelve five-membered rings [43], the corners of which cannot be occupied by alternating boron and nitrogen atoms.

consequence, its bulk modulus would, according to (1.24), be lower than that of diamond despite the shorter bond length[2].

For this reason, only tetrahedrally bonded compounds in the B/C/N system are considered in this book[3]. Figure 1.3 shows that diamond is the only elemental material (other than the – amorphous – ta-C) and c-BN the only binary compound of this kind that has been realized up to now (B_4C is hard but not superhard). For the binary compound β-C_3N_4, it was proposed by Liu and Cohen [22] that this material should be superhard if it is possible to realize it. In recent times, the preparation ofternary compounds such as BC_2N and BC_4N has been suggested [36, 38, 49, 50, 51, 52], which should also be superhard and possess further interesting properties. Nevertheless, the deposition of such materials in form of thin films has been scarcely investigated and has not been realized yet. For these reasons this book concentrates on carbon (diamond and ta-C) and the binary compounds c-BN and β-C_3N_4.

2.2.2 Carbon and Boron Nitride

Boron nitride is isoelectronic to carbon; the B–N and B=N groups have chemical properties similar to those of their carbon analogs C–C and C=C, respectively. In fact, a great variety of organic molecules can be realized with B–N instead of C–C groups [39, 53]. It is therefore not surprising that great similarities exist also between the major crystalline modifications of the two materials.

Diamond and c-BN are the most important crystalline tetrahedral (i.e. sp^3-bonded) carbon and BN modifications. Both crystallize in the zinc blende structure and have very similar lattice constants (Table 2.1). The crystal structures of the major sp^2-bonded phases, graphite and h-BN are also almost identical: both materials have a layered structure; the individual layers are composed of six-membered rings but between these layers only weak van der Waals interactions take place. There are nevertheless slight differences between graphite and h-BN with respect to the stacking of the layers.

Under normal conditions, graphite is the stable crystalline modification of carbon, whereas diamond is metastable, as shown in the phase diagram in Fig. 2.1. Accordingly, high-pressure/high-temperature (HPHT) processes are used to produce diamond by transformation of graphite. The high temperatures are, however, only necessary for kinetic reasons (diffusion and reaction velocities; see Fig. 2.1).

---

[2] Equation (1.24) yields for H6-C a value of $K = 390$ GPa, in good agreement with the value of 372 GPa [48] obtained by ab initio calculations [48]. Up to now, it has not been possible to realize H6 carbon; according to new calculations, this modification seems to be not stable relative to diamond [38, 48].

[3] In β-C_3N_4, the nitrogen atoms are bonded trigonally. Some new calculations [27, 32] show, however, that C_3N_4 modifications with tetrahedrally bonded nitrogen may also be stable; see Sect. 2.4.4.

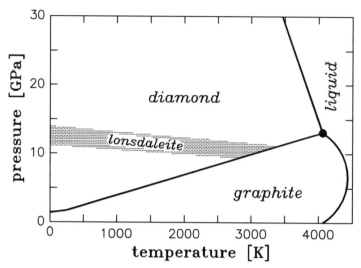

Fig. 2.1. Phase diagram of carbon. The term 'lonsdaleite' labels sp^3-bonded hexagonal carbon (hexagonal diamond)

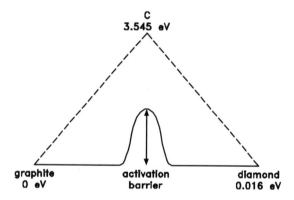

Fig. 2.2. Energy diagram of carbon according to [54]

However, diamond is only slightly unstable with respect to graphite, as shown in the schematic energy diagram in Fig. 2.2 [54, 55]. On the other hand, there exists a high activation barrier between them, the exact height of which is nevertheless unknown. The small energy difference on the one hand and the high activation energy on the other hand imply two conclusions which are of importance for the deposition of diamond films. First, there is a certain probability that during a deposition process both materials, diamond and graphite, are obtained; second, once diamond has been deposited, a re-transformation into graphite is extremely improbable under normal conditions.

Despite all analogies with respect to crystal structure and bond length between diamond and c-BN on the one hand, and graphite and h-BN on the other hand, which result in a large variety of common or similar physical

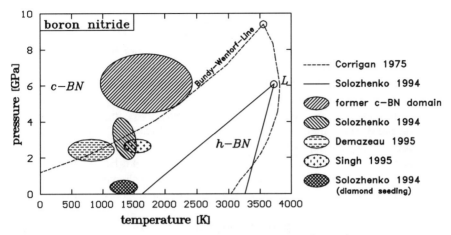

Fig. 2.3. Phase diagram of boron nitride according to [52, 56]. The *dashed line* marks the c-BN domain proposed by Corrigan and Bundy [57]; the *solid line* is based on new calculations by Solozhenko [56]. The *hatched areas* mark successful HPHT synthesis experiments, according to [52, 56, 58]

properties of the corresponding modifications (Table 2.1), there are nevertheless some basic differences between the boron nitride system and the carbon system.

1. Graphite is the thermodynamically stable crystalline carbon modification under normal conditions (Fig. 2.1), whereas diamond is metastable. Until very recently it has been assumed that the relations are similar for boron nitride, i.e. h-BN is stable and c-BN is metastable (see the 'Bundy–Wentorf line' in Fig. 2.3). This assumption was based based on an extrapolation of data obtained in a limited parameter space, and on considerations of the analogy to the carbon system [57]. Indeed, one can produce c-BN – as in the case of carbon – by transformation of h-BN under HPHT conditions [59, 60, 61].

   New calculations [52, 56] and, especially, new experimental results [56, 58, 62], however, make it very probable that under normal conditions c-BN is the stable modification while h-BN is metastable (Fig. 2.3). Nevertheless, this question is still controversial in the literature[4]. Irrespective of the stability of the two modifications, in every case there exists – as in the case of diamond – a kinetic barrier between h-BN and c-BN, preventing a transformation between the two phases under normal conditions.

2. The carbon–carbon bond is purely covalent, whereas the boron–nitrogen bond is partially ionic (Table 2.1). For the sp^2 modifications graphite and h-BN this implies, for example, that the π electrons in the case of graphite

---

[4] For the latest developments in this area see [63] and the literature cited therein.

are delocalized within the six-membered rings, whereas in the case of h-BN they are localized mainly at the nitrogen atoms. Graphite is therefore electrically conductive normal to the c axis, but not h-BN. Further, the ionicity of the B–N bond causes the above in the stacking of the six-membered rings between h-BN and graphite previously mentioned.

In the context of the topic of this book, the following consequences of the partially ionic character of the B–N bond are of importance. On the one hand, the ionicity of course influences the chemical processes taking place during the deposition of c-BN films[5]. On the other hand the ionicity causes, according to (1.24), a lower bulk modulus of c-BN (Table 2.1) and thus a lower hardness, too. Nevertheless, it will be shown in Sect. 3.3 that the ionicity of the B–N bond is of advantage for the crystallinity of c-BN films.

3. Boron nitride is a binary compound. This implies the following differences from carbon:

   (i) The deposition of c-BN films requires one to fulfill the stoichiometry condition $B/N \approx 1$. In comparison to the carbon system, this implies an additional constraint.
   (ii) The growth of BN films requires the attachment of boron atoms to nitrogen atoms and vice versa. If the formation of B–B or N–N bonds is favored for some reason, this can cause the termination of the crystal growth[6].
   (iii) The low-index crystal planes (e.g. {100}, {111}) of boron nitride can be either boron- or nitrogen-terminated. This complicates crystal growth by the attachment of well-defined species, which indeed takes place during diamond deposition as will be shown in Sect. 3.2.

4. None of the crystalline boron nitride modifications exists naturally [61].
5. Whereas in the carbon system also an amorphous, mainly sp^3-bonded modification exists in addition to diamond (ta-C), the preparation of an amorphous, tetrahedral, stoichiometric BN compound has proven impossible up to now; according to molecular-dynamics (MD) calculations, the deposition of such structures should always lead to agglomerations of boron [67][7].

---

[5] In general, isoelectronic boron nitride and carbon compounds (e.g. borazine ($B_3N_3H_6$) and benzene (C_6H_6), and amino boranes ($R_2B=NR_2$) and the analogous C=C compounds) possess similar physical properties but differ distinctly with respect to their chemical reactivity; usually the BN compounds are considerably more reactive than their carbon analogs [39]. This is mainly caused by the polarity of the bond.

[6] This leads to the fact that HPHT c-BN possesses considerably lower crystal sizes and a considerably poorer quality than HPHT diamond [52, 64]. For the state of the art in c-BN HPHT growth see [65, 66].

[7] The deposition of boron-rich, amorphous, mainly sp^3-bonded layers, on the other hand, is possible without problems [68], in agreement with the MD calculations.

It will be shown in Chaps. 3 and 4 that great differences exist between the techniques successfully applied for the deposition of diamond and c-BN films. The deposition of c-BN is not possible with the methods which are best suited for diamond synthesis, and vice versa. To what extent this arises from the differences in the physical and chemical properties of diamond and c-BN listed above is the main topic of this book. Further, taking into account that besides diamond and c-BN further superhard materials (may) exist in the B/C/N system (ta-C, β-C_3N_4 and ternary compounds), the following, more general question arises which has already been raised by Yarbrough [69]:

> Is the low pressure synthesis of diamond simply a peculiarity of the (hydro) carbon chemistry with no analogous processes existing in other systems, or can the mechanisms of the diamond synthesis be applied in order to prepare new materials with other structures and/or compositions which are of interest for research and applications?

At the end of this book it will be shown that, according to the present state of knowledge, the answer to this question is that to a large extent the low-pressure synthesis of diamond is indeed a peculiarity of the carbon system not easily transferable to other systems.

2.3 Properties of Diamond-Like Materials

The short bond lengths (and the corresponding high atomic density), the high binding energy and the low ionicity of the compounds in the B/C/N system are responsible not only for a high bulk modulus and thus a high hardness, but also for a variety of other interesting material properties [26]. These are discussed in the following sections, before the state of the art concerning the deposition of the four materials considered in this book in the form of thin films is briefly summarized in a short overview at the end of this chapter.

Table 2.1 summarizes some important properties of the diamond-like materials considered in this book and, for comparison purposes, also of the two sp^2 modifications graphite and h-BN. This table and the discussion in the following sections show that the bond characteristics which define this class of material and which are listed in the first group in Table 2.1 are responsible not only for the extreme mechanical properties summarized in the second group but also the extreme thermal and vibrational properties of the third group. The fourth group, finally, shows that there nevertheless exist important differences with respect to the optical and electronic characteristics, which are of special interest for the design of a material.

Table 2.1. Properties of the four materials considered in this book. For comparison purposes, the sp² modifications graphite and h-BN are also included. For these hexagonal materials, in the case of anisotropic properties first the value for the a axis and then that for the c axis is given. The first group of entries lists those properties which, according to (1.24) and (1.19), determine the hardness either directly or via the bulk modulus; the second contains the mechanical and elastic characteristics. The third group lists properties which follow immediately from the short bond length and high atomic density. The fourth group includes some further features which are of interest for applications. It should be remembered that β-C_3N_4 is a hypothetical material; the properties listed therefore are based mainly on calculations; further, that the term ta-C labels not a well-defined material but rather a class of coatings, which in addition are not crystalline but amorphous.

Sources: For the mechanical and elastic properties of diamond and c-BN [13, 70]; for the properties of β-C_3N_4 [31, 71]; in all other cases [26, 72, 73] and standard data collections

Property	Diamond	ta-C	c-BN	β-C_3N_4	Graphite	h-BN
Bond length (nm)	0.154	$\approx$ 0.152	0.157	0.147	0.141	0.145
Bond length (c) (nm)	—	—	—	—	0.335	0.333
Coordination number	4	3–4	4	3.43	3	3
Ionicity according to [2]	0	0	1	0.5	0	1
Ionicity according to (1.21)	0	0	0.25	0.08	0	0.25
Atomic density (nm^{-3})	176.3	$\approx$ 150	170	162.5	113.9	115.8
Density (g/cm^3)	3.51	> 3	3.47	3.54	2.27	2.27
Bulk modulus^a (GPa)	443	—	369	427		
Bulk modulus^b (GPa)	435	—	367	414–438	—	—
Young's modulus (GPa)	1140	—	712	—	18.8/5.2	87/34
Poisson's ratio	0.07	—	0.09	—		0.3
Vickers hardness (GPa)	100	> 40	$\approx$ 70	—		$\approx$ 10
Debye temperature (K)	1860	—	1700		760	598
Sound velocity (10^4 m/s)	1.1–1.8	—	$\approx$ 1.5	$\approx$ 1.1		
Melting point (K)	3800	—	> 2973	—	4000^c	2600^d
Thermal conductivity^e (W/mK)	2000	100–700	1300	High	10	70/0.8
Thermal expansion (10^{-6})	0.8	< 5	4.8	—		2.7/3.7
Band gap (eV)	5.45	$\approx$ 3	> 6	6.4	Metallic	4.5
Resistivity (Ω cm)	> 10^{16}	10^{10}	> 10^{16}	—	10^{-3}/10^0	10^{10}/10^{12}
Refractive index^f	2.4	2–3	2.12	—		2.10/1.75

^{a}Experimental data (for β-C_3N_4: ab initio calculation); ^{b}according to (1.24); ^{c}sublimation temperature; ^{d}decomposition temperature; ^{e}at 293 K; ^{f}at 600 nm.

2.3.1 Thermal and Vibrational Properties

Melting Point and Thermal Expansion. All diamond-like materials possess a high melting point T_m and a low linear thermal expansion coefficient α. The starting point of the discussion [74] is the dependence of the interatomic potential on the distance between the atoms shown in Fig. 1.1. This potential consists of an attractive term $-E_A r^{-n}$ and a repulsive term $E_R r^{-m}$:

$$E(r) = E_R r^{-m} - E_A r^{-n} . \tag{2.1}$$

The lattice energy is given by the energy at the equilibrium distance $E_0 = E(r_0)$; at r_0, the force $F(r) = E'(r)$ vanishes:

$$F(r_0) = E'(r_0) = -\frac{m}{r_0} E_R r_0^{-m} + \frac{n}{r_0} E_A r_0^{-n} = 0 . \tag{2.2}$$

From this equation, the repulsive and attractive energy components at r_0 can be expressed in terms of the lattice energy E_0:

$$E_R r_0^{-m} = E_0 \frac{n}{m-n} , \qquad E_A r_0^{-n} = E_0 \frac{m}{m-n} . \tag{2.3}$$

The Young's modulus $\mathcal{E}$ is in turn given by the derivative of the force/distance curve, i.e. the curvature of the potential curve at r_0:

$$\mathcal{E} = \frac{1}{r_0} F'(r_0) = \frac{1}{r_0} E''(r_0) = mn \frac{E_0}{r_0^3} . \tag{2.4}$$

The third derivative of $E(r)$ describes the asymmetry of the potential curve:

$$E'''(r_0) = -\frac{nm(m+n+3)}{r_0^3} E_0 . \tag{2.5}$$

Thus, the potential curve in the vicinity of the minimum can be expressed as a Taylor series

$$\begin{aligned} E &= E_0 + \frac{1}{2} E''(r_0) x^2 + \frac{1}{6} E'''(r_0) x^3 \\ &= E_0 \left(1 + \frac{1}{2} \frac{nm}{r_0^2} x^2 + \frac{1}{6} \frac{nm(n+m+3)}{r_0^3} x^3 \right) , \end{aligned} \tag{2.6}$$

where $x = r - r_0$ is the distance from the minimum.

At a given temperature T, an atom vibrates with a potential energy $(3/2) kT$ above the potential minimum. Owing to the asymmetry of the potential, the center of the vibration moves from r_0 to a larger distance by

$$\delta = \frac{(1/6) E'''(r_0)}{(1/2) E''(r_0)} \frac{3}{2} kT = \frac{m+n+3}{2mn} \frac{kT}{E_0} r_0 . \tag{2.7}$$

For the linear thermal expansion coefficient, we thus obtain

$$\alpha = \frac{\delta}{r_0} \frac{1}{T} = \frac{m+n+3}{2mn} \frac{k}{E_0} . \tag{2.8}$$

A crystal starts to melt if the amplitude of the vibrations reaches a certain percentage of the bond length [8] or kT reaches a certain percentage of the binding energy E_0 [74]. In both cases it follows that $T_\mathrm{m} \propto E_0$. The exponents n, m in (2.1), on the other hand, are constant for a given class of materials[8]. As a result, the thermal expansion coefficient α is inversely proportional to E_0 within a class of materials. Finally, the Young's modulus is, according to (2.4), higher the higher the binding energy and the lower the bond length are. Since $\mathcal{E}$ is correlated with the bulk modulus via

$$K = \frac{\mathcal{E}}{3(1-2\mu)}, \qquad (2.9)$$

and since all covalent materials possess a low Poisson's ratio [8], one finds in summary, that, owing to their high binding energy and short bond length, the covalent diamond-like materials considered in this book have a high hardness, a high melting point and a low thermal expansion coefficient in common.

Velocity of Sound. The velocity of sound v_s in an infinite solid is determined by the Young's modulus $\mathcal{E}$, the Poisson's ratio μ and the density ρ [75][9]:

$$v_\mathrm{s} = \sqrt{\frac{\mathcal{E}}{\rho} \frac{1-\mu}{(1+\mu)(1-2\mu)}}. \qquad (2.10)$$

The Young's modulus is correlated via $\mathcal{E} = 3(1-2\mu)K$ with the bulk modulus, which in turn is, according to Cohen's relation, extremely high for diamond-like materials, which possess in addition a very low Poisson' ratio (Table 2.1) like all covalent materials [8]. Their mass density ρ, on the other hand, is low despite the high atomic density. As a consequence, diamond-like materials are marked by a high velocity of sound, which correlates with the bulk modulus (and thus with the hardness).

Debye Temperature. According to Debye's theory, for each solid there exists a maximum phonon frequency ω_D which is characteristic of the material:

$$\omega_\mathrm{D} = v_\mathrm{s} \sqrt[3]{\frac{6\pi^2 N}{V}} = v_\mathrm{s} \sqrt[3]{6\pi^2 n_\mathrm{a}}. \qquad (2.11)$$

Here, v_s is the velocity of sound and N is the number of atoms in the volume V. This maximum frequency can be related via

$$\Theta = \frac{\hbar \omega_\mathrm{D}}{k_\mathrm{B}} = \frac{\hbar}{k_\mathrm{B}} v_\mathrm{s} \sqrt[3]{6\pi^2 n_\mathrm{a}}, \qquad (2.12)$$

where k_B is Boltzmann's constant, to a characteristic temperature or Debye temperature Θ; T/Θ is therefore a measure of the degree of excitation of the phonon spectrum.

---

[8] For ionic, covalent and metallic solids, $n = 1, 2$ and 4, respectively; m is typically in the order of 10 [7].

[9] For a rod–shaped body 2.10 reduces to $v_\mathrm{s} = \sqrt{\mathcal{E}/\rho}$, since in contrast to an infinite solid, transverse contractions are possible.

Equation (2.12) shows that Θ is determined by the velocity of sound and the atomic density. Both v_s and n_a are very high for diamond-like materials; as a consequence, their Debye temperatures are high, too. Indeed, diamond possesses the highest Debye temperature of all known materials (Table 2.1). However, although a high Θ is of advantage for the properties of a material, as will be shown in the remainder of this section, it is of great disadvantage from a technological point of view: the temperature required for the relaxation of defects created during the deposition of a material or its processing, by means of annealing, is typically in the order of Θ or even higher. For diamond-like materials, therefore, post-deposition/post-processing improvement of the crystal quality by annealing processes is difficult or even impossible [76].

Thermal Conductivity. The transport of heat in solids is carried out either by electrons or by phonons; for the covalent materials considered here the phonon contribution dominates. In this case the thermal conductivity can be expressed, according to the Debye equation, as [77, 78]

$$\mathcal{K} = \frac{1}{3} C_p v_s \lambda \,, \tag{2.13}$$

where C_p is the heat capacity and λ the mean free path of the phonons. Thus, for a high thermal conductivity $\mathcal{K}$, a high velocity of sound v_s and a large mean free path are required.

The velocity of sound is, as shown above, high for the diamond-like materials considered here. The mean free path of phonons, on the other hand, is determined by scattering processes either with other phonons or at lattice defects and surfaces. Phonon-scattering processes of the form

$$\boldsymbol{k}_3 = \boldsymbol{k}_1 + \boldsymbol{k}_2 + \boldsymbol{G} \tag{2.14}$$

($\boldsymbol{G}$ being a reciprocal-lattice vector) can contribute to the thermal resistance only if $\boldsymbol{G} \neq 0$ (umklapp processes). The probability of umklapp processes, on the other hand, is determined by the Debye temperature:

$$\lambda_{\text{Umklapp}} \propto \exp\left(-\frac{\Theta}{2T}\right) \,; \tag{2.15}$$

as a consequence, they dominate at higher temperatures. Since the Debye temperature of the diamond-like materials is very high, as we have seen, umklapp processes occur only at temperatures much higher than those required for other materials and yield, even at room temperature, a much smaller contribution.

Table 2.2 demonstrates the influence of both effects using diamond as an example: at 80 K, umklapp processes can be neglected; scattering of phonons at defects is responsible for the difference between the almost defect-free type IIa diamond and type Ia diamond, which contains nitrogen as an impurity[10].

---

[10] By elimination of the natural concentration of 1 % of ^{13}C atoms, $\mathcal{K}$ can be increased even further (to approximately 3300 W/m K at room temperature for 0.07 % of ^{13}C) [78, 79].

Table 2.2. Thermal conductivity $\mathcal{K}$ of diamond in units of $\mathrm{W\,m^{-1}\,K^{-1}}$ according to [70]

Type	293 K	80 K
Ia	600–1000	2000–4000
IIa	2000–2200	$\leq 15\,000$

The decrease of $\mathcal{K}$ when switching to room temperature is caused by the increasing influence of umklapp processes. Taking all together, nevertheless, the thermal conductivity of diamond-like materials is, despite the missing contribution of electrons, much higher even at room temperatures than it is for other solids, because of the high velocity of sound and the large mean free path of phonons (for type IIa diamond, $\mathcal{K}$ is approximately five times higher than for copper or silver [79]).

Plasmon Frequency. The plasmon frequency of a solid is basically determined by the electron concentration n_e:

$$\omega_\mathrm{p} = \sqrt{\frac{n_\mathrm{e} e^2}{\epsilon_0 m_\mathrm{e}}}, \tag{2.16}$$

where m_e is the electron mass. The electron density is in turn determined by the atom density n_a; for the covalent materials considered in this section, $n_\mathrm{e} = 4 n_\mathrm{a}$ [80]. Since all diamond-like materials possess a short bond length and thus a high atomic density (Table 2.1), they are also distinguished by a high plasmon frequency [26].

2.3.2 Optical and Electronic Properties

On the other hand, there are some important material properties which do not correlate directly with hardness. This is especially true for optical and electronic characteristics. In most cases the reason can be found in the fact that the bond length d and the ionicity a_i cause opposite effects. This means that for the purely covalent semiconductors diamond, silicon and germanium the optical and electronic properties often correlate with the bond length and thus with hardness; the correlation is lost, however, if partially ionic compounds such as the III/V semiconductors (here, especially, c-BN) are taken into account. As examples, the band gap E_G and the refractive index n are discussed in the following (see Table 2.3).

Band Gap. According to the dielectric model of semiconductors of Phillips and van Vechten [80, 81], the average band gap E_g, defined as the difference between the average energies of the conduction band and valence band,

$$E_\mathrm{g} = \bar{E}_\mathrm{c} - \bar{E}_\mathrm{v}, \tag{2.17}$$

Table 2.3. Electronic and dielectric parameters, and refractive indices in the visible for the covalent semiconductors C, Si and Ge, and the analogous III/V semiconductors [81, 82]

Material	E_G (eV)	E_g (eV)	E_h (eV)	C (eV)	$\hbar\omega_p$ (eV)	$\epsilon(0)$	$\epsilon(\infty)$	$n^\dagger$
C	5.45	13.5	13.5	0	31.2	5.66	5.66	2.41
Si	1.12	4.77	4.77	0	16.6	12.0	12.0	3.88
Ge	0.64	4.31	4.31	0	15.6	16.0	16.0	4.12
c-BN	> 6	15.2	13.1	7.71	30.6	7.1	4.5	2.12
AlP		6.03	4.72	3.76	16.5		6.0	2.85
GaAs	1.4	5.20	4.32	2.90	15.6	13.0	5.2	3.86

† at 630 nm.

consists of a covalent contribution E_h and an ionic contribution C:

$$E_g^2 = E_h^2 + C^2 . \tag{2.18}$$

The covalent gap E_h is determined by the bond length d^{11}:

$$E_h \propto d^{-2.5} . \tag{2.19}$$

This means that for purely covalent materials ($C = 0$) the average band gap correlates via the bond length with the hardness (see Table 2.3)[12].

The ionic contribution C to the average band gap can, according to Phillips [80, 81], be described by the covalent radii r_A, r_B and the atomic numbers Z_A, Z_B of the participating elements (with $r_A + r_B = d$):

$$C \propto \left(\frac{Z_A}{r_A} - \frac{Z_B}{r_B} \right) e^{-ik_s d} \tag{2.20}$$

(k_s is the Thomas–Fermi screening length). Thus, the contribution of C is the higher the larger the difference between the electronegativities of the participating elements; in good approximation, it is given by

$$C(\text{eV}) = 5.75(\epsilon_B - \epsilon_A) . \tag{2.21}$$

As a consequence of these dependences, the contribution of C to the average band gap within a period is the higher the further the elements are separated in the periodic table (compare the values of $C = 14.7$ eV and $E_g = 18.8$ eV for BeO with the data in Table 2.3); for a given type of semiconductor, on the other hand, C decreases with increasing period. Taking everything together,

---

[11] In this book, the influence of the d electrons, which has to be taken into account here and in the following discussion for elements from the third and higher periods [81], will be neglected since the aim of this section is only to outline some general trends; moreover, the main topic is the elements of the first row.

[12] Indeed, (2.19) is the starting point of Cohen's derivation of the $d^{-3.5}$ dependence of the bulk modulus briefly mentioned in Sect. 1.4.4.

therefore, the ionic contribution to the average band gap has the effect that E_g is a function of the bond length as well as of the ionicity; since their influences on E_g are opposite, there is no correlation of the average band gap with the hardness.

The quantity that is decisive for the electrical properties of a semiconductor, however, is not the average band gap E_g but the band gap E_G, defined as the difference between the conduction band minimum and valence band maximum:

$$E_G = E_c - E_v .\qquad(2.22)$$

The energy bands of a semiconductor are a function of the crystal direction. Their explicit shapes depend in a complex manner on the structure of the material; as a consequence, E_c and E_v do not necessarily occur at the same wave vector. It is therefore more difficult to derive general expressions for E_G. Nevertheless, E_c, E_v and E_G follow in general the trends outlined above for E_g [81], as is also evident from the data in Table 2.3.

Refractive Index. The refractive index n in the visible part of the spectrum is approximately given by the square root of the optical dielectric constant:

$$n \approx \sqrt{\epsilon(\infty)} ;\qquad(2.23)$$

$\epsilon(\infty)$ is related to the static dielectric constant $\epsilon(0)$ via the dielectric function or the Lyddane–Teller–Sachs equation:

$$\epsilon(\infty) = \epsilon(0) - \frac{4\pi e^2 n_{AB}}{m_{AB}\Omega_{TO}} \quad \text{or} \quad \frac{\epsilon(\infty)}{\epsilon(0)} = \frac{\Omega_{TO}^2}{\Omega_{LO}^2} .\qquad(2.24)$$

Here, n_{AB} and m_{AB} are the concentration of heteropolar pairs and their reduced mass, respectively; Ω_{TO} and Ω_{LO} are the frequencies of the transverse and longitudinal phonons in the center of the Brillouin zone. For purely covalent (homopolar) materials, $n_{AB} = 0$ and $\Omega_{TO} = \Omega_{LO}$; thus, the refractive index is given by the static dielectric constant $\epsilon(0)$; $\epsilon(0)$ in turn depends, according to Phillips [81], on the plasmon frequency ω_p and the average band gap E_g:

$$\epsilon(0) = 1 + \frac{(hw_p)^2}{E_g^2}\left(1 - \frac{E_g}{4E_f}\right) ,\qquad(2.25)$$

where E_f is the width of the valence band. The plasmon frequency scales according to (2.16), with $d^{-1.5}$; the average band gap, on the other hand, scales according to (2.19), with $d^{-2.5}$; as a consequence, the refractive index of the purely covalent semiconductors increases with increasing bond length and thus correlates inversely with the hardness.

Given the same static dielectric constant, $\epsilon(\infty)$ and thus also n are, for a polar compound, smaller than for a purely covalent semiconductor, according to (2.24). The difference term contains, among others, the bond length (via n_{AB}), the reduced mass of the elements of the compound and the interatomic force constant (via Ω_{TO}); as a consequence there is no general trend for these materials, as is also evident from Table 2.3.

Conclusions. The fact that not all properties of the diamond-like materials are correlated with the hardness and density (as shown above, the most important of these are the optical and electronic properties) is of major importance from a technological point of view as well as from the point of view of fundamental physics. On the one hand, it opens up the possibility of material design: if it is possible to realize, in addition to the members of this class already known and considered in this book, new compounds, e.g. β-C_3N_4 or ternary compounds of the type $B_xC_yN_z$, this opens up the possibility to tailor superhard materials with all of the extreme properties discussed above but variable optical and electronic properties. Furthermore, new members of the class of diamond-like materials should also provide some new insight into the dependence of the optical and electronic properties on the structure and composition of these compounds.

2.4 State of the Art

In the following sections, the state of the art concerning the deposition techniques that have been applied, the properties of the films obtained so far, and the current actual problems is briefly outlined for the four superhard materials considered in this book. For details, we refer to the appropriate sections in Chaps. 3 and 4.

2.4.1 Diamond

Carbon can exist in a vast variety of crystalline and, especially, amorphous modifications. Besides diamond and graphite, the most important crystalline phase is lonsdaleite (hexagonal diamond), which is distinguished from diamond only by the different stacking of the crystal planes [83] and which has properties very close to those of diamond. The amorphous modifications, some of which possess extreme hardness (e.g. ta-C and DLC), will be discussed in detail in Sect. 2.4.2.

This variety of possible carbon modifications led during the 1980s to great problems concerning nomenclature: which coatings should be labeled as 'diamond'? These problems in turn rendered comparisons of different papers published in the literature extremely difficult. In order to solve this problem, Messier et al. [33] proposed in 1987 a 'working definition' for diamond films which is now generally accepted; according to this definition, diamond films have to fulfill the following requirements:

- scanning electron microscopy images show well-faceted crystals
- X-ray or electron diffraction patterns show the reflections typical of diamond
- Raman spectra show the typical diamond line at 1332 cm^{-1}.

Comparing the four superhard materials discussed in this book, the state of the art is by far most advanced in the case of diamond. This is true for technological aspects as well as from the point of view of the theoretical understanding of the fundamental mechanisms of diamond growth. This reflects the relatively early start of intensive research in the field of diamond deposition (beginning of the 1980s)[13] as well as the vast number of research teams world-wide working in this area.

As a consequence, most of the problems being currently investigated are related to the application of diamond films. The deposition techniques and basic characterization methods are, meanwhile, well established [84, 85, 86, 87]. In the field of modeling, at least a fundamental understanding of the basic processes has been obtained [55, 76, 87, 88], even if the detailed mechanisms are still controversial (see Sect. 3.2.4) and, as a consequence, are being investigated intensively. Nevertheless, at the present time research in the area of low-pressure diamond synthesis is focused on the optimization of the deposition techniques and of the properties of the diamond films for future applications. Examples of the current topics [89] with respect to process optimization are investigations aimed at increasing the growth rates, lowering the substrate temperatures and reducing the costs of diamond deposition; investigations regarding special applications include reduction of the surface roughness (optics), increase of the thermal conductivity, reduction of the defect density (optics and electronics) and, especially, the problem of heteroepitaxy (electronics), to name but a few.

2.4.2 Tetrahedral Amorphous Carbon

The number of possible amorphous carbon films, considering the nature of bonding, composition and structure, is immense. This is especially true if the consideration also includes hydrogen containing films and if, in addition, classical polymers such as polyacetylene $(-CH-)_n$ and polyethylene $(-CH_2-)_n$ are taken into account. This variety of possible structures corresponds to a extreme variation of film properties, which ranges from soft to superhard. A classification of the various types of (hydro)carbon layer can be performed through their hydrogen content, on the one hand, and their density [83] or the fraction of sp^3 bonds (Fig. 2.4), on the other hand.

The solid line in Fig. 2.4 represents the dependence of the sp^3 content on the hydrogen concentration for a completely 'random covalent network' according to the model of Angus and Jansen [90]. In this case, the number of constraints per atom equals the number of mechanical degrees of freedom per atom. For higher hydrogen concentrations, the films possess more degrees of freedom than constraints and are therefore soft[14]. In films with lower H

---

[13] Research on c-BN started at the end of the 1980s; for β-C_3N_4 it started at the beginning of the 1990s.

[14] For even higher hydrogen contents, no stable films are possible.

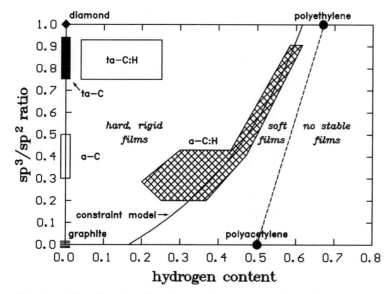

Fig. 2.4. Classification of carbon and hydrocarbon films (schematic). This diagram summarizes similar presentations in [72, 26, 83, 90]

concentrations, on the other hand, the number of constraints exceeds the number of degrees of freedom; as a consequence, such films are hard but also possess stress. As is evident from Fig. 2.4, there is a class of films (amorphous carbon, a-C:H) the composition of which satisfies the conditions for a random covalent network. In addition, however, there are also films with a lower or even vanishing hydrogen content, which – depending on the sp^3 content – are labeled as amorphous carbon (a-C) or tetrahedral amorphous carbon (ta-C). According to the model of Angus and Jansen, such films are hard or even superhard, but also possess stresses (in some cases very large).

This variety of possible amorphous carbon and hydrocarbon films thus calls for unequivocal definitions of the various modifications in order to allow comparison of different papers in the literature and to avoid confusion[15]. As proposed in previous publications [72, 91], in this book amorphous carbon films will be labeled as 'ta-C' only if the following requirements are fulfilled:

- the sp^3 content of the carbon skeleton amounts to at least 60 %
- the density is greater than 2.5 $g\,cm^{-3}$, typically reaching values even greater than 3 $g\,cm^{-3}$
- the optical band gap is at least 1.5 eV.

---

[15] Unfortunately, the nomenclature used in the literature is far from systematic. In particular, the term 'diamond-like carbon' (DLC) is used indiscriminately for ta-C, a-C and a-C:H films.

Besides the (almost) hydrogen-free ta-C, films with very high sp^3 contents exist which have hydrogen contents up to 25% (ta-C:H, see Fig. 2.4) [92]. Their properties are very similar to those of ta-C [72]; in the remainder of this book, these films will be discussed together with ta-C layers.

The number of theoretical and experimental papers published in the literature concerning the deposition of ta-C – in the strong sence of the above definition – is lower by far than the number of publications on the deposition of diamond or c-BN. Accordingly, the state of the art is distinctly less advanced than that for diamond technology; it corresponds roughly to that for c-BN deposition (see below). The necessary deposition and characterization techniques are now well established; industrially relevant applications, however, have not been developed so far, thus restricting the work performed to the laboratory scale. This is mainly due to the properties of the films obtained up to now: ta-C layers possess, on the one hand, a high hardness and other excellent mechanical properties but also, on the other hand, a high compressive stress which prevents the deposition of thicker films. Further current problems concern the poor temperature stability and the incorporation of macroparticles (see Sect. 3.4).

2.4.3 Cubic Boron Nitride

As in the case of carbon, a vast variety of crystalline and amorphous modifications of boron nitride are known. In analogy to lonsdaleite, hexagonal wurtzite BN (w-BN) forms a second crystalline sp^3 phase, and, similar, sp^2-bonded rhombohedral BN (r-BN) is distinguished from h-BN only by the stacking of the individual layers [93, 94]. Turbostratic (sp^2-bonded) BN forms an intermediate between crystalline and amorphous compounds: the six-membered BN rings are arranged in stackings, and there is indeed a short-range order; on a macroscopic scale, however, the direction of the c axis shows a random distribution. Finally, a variety of amorphous sp^2 BN compounds exist which are mainly distinguished by their density; in contrast to their carbon analogs, however, no classification has been introduced up to now. It is very interesting and of importance for the modeling of c-BN deposition that – in contrast to carbon (ta-C) – no stoichiometric amorphous sp^3 modification seems to exist for boron nitride.

The variety of possible BN modification causes – as in the case of carbon – difficulties in the unequivocal identification of c-BN. These difficulties are aggravated by the fact that boron nitride is a binary compound, which means that the stoichiometry can vary. In addition there is – in contrast to diamond – the possibility of the incorporation of impurities[16]. Both effects can, as will be shown in Sect. 3.3.1, prevent the formation of c-BN. In analogy to the

---

[16] The most important of these are – as for all processes in thin-film deposition – cargon and oxygen (Sect. 3.3.1). In diamond deposition, the presence of these elements is at least not a major disadvantage: carbon contamination (e.g. from the walls of the deposition reactor or from pumping oil) will be transferred to

definition of diamond by Messier et al. [33], we have therefore proposed a 'working definition' for c-BN [72, 96] according to which c-BN films have to fulfill the following requirements:

- The infrared spectra of c-BN layers show only the c-BN reststrahlen band at 1065 cm^{-1} and the bands of the h-BN stretching vibration at 1380 cm^{-1} and the h-BN bending vibration at 800 cm^{-1}.[17]
- Diffraction spectra of c-BN layers show the reflections assigned to c-BN. The films are nanocrystalline with crystallite sizes between 5 and 100 nm.
- c-BN films have an almost stoichiometric composition ($0.9 \leq$ BN ≤ 1.1). The concentrations of impurities, especially oxygen, are low.

Compared to the low-pressure synthesis of diamond, the state of the art of the deposition of cubic boron nitride is approximately ten years behind. This is especially true for the technological aspects of c-BN deposition, in contrast to the understanding of the basic mechanisms of the underlying processes. With respect to the modeling of c-BN deposition, an intense controversy is indeed taking place at the present time[18]. Nevertheless, all models presently discussed share some common basic aspects; the arguments concern, rather, the weights of the influences of various effects (compare Sect. 3.3.4). Likewise, the coating techniques suitable for c-BN deposition are now well-established, even if these techniques are at present almost exclusively applied on a laboratory scale. The great problem of c-BN deposition at the present time lies in the basic properties of the deposited layers, which fail to fulfill technological requirements by far. The relatively low growth rates and the nanocrystalline structure of the films are aspects which could be improved or, at least for certain applications, even tolerated. This does not hold, however, for the two major problems of c-BN deposition: the high compressive stress and the extremely poor adhesion, which at present render the deposition of thicker films ($>$ some hundred nm) with high c-BN contents impossible. These problems prevent not only the technological application of c-BN coatings but also their characterization with respect to technologically revelant properties. As a consequence, current research in the field of c-BN deposition concentrates mainly on possible solutions of this stress/adhesion problem.

---

diamond; the presence of oxygen improves the quality of the deposited diamond films (Sect. 3.2.2). Under standard conditions, there is a significant probability of incorporation into the diamond lattice only for nitrogen (however, with a very low probability) [95] and hydrogen.

[17] It is difficult to distinguish the different sp^2-bonded BN materials such as h-BN, t-BN and boron-rich a-BN by means of IR spectra only. As long as only sp^2 contamints in c-BN films are considered, all these modifications are labeled as h-BN for simplicity in the following. The only exception will be the discussion of the nucleation of c-BN in Sect. 4.4.

[18] In a recent publication [97], we presented a review of the different models proposed in the literature and the main points of discussion.

2.4.4 β-C_3N_4

It is a misunderstanding widespread in the literature that Liu and Cohen predicted in their papers [2, 22, 38] the *existence* of the superhard material β-C_3N_4. Rather, Liu and Cohen start from (1.24): superhard materials are marked by a short bond length and a low ionicity. Such materials are – as discussed above – to be found in the periodic table in the vicinity of carbon; one of the possible materials – which has not been realized up to now – is carbon nitride. The prediction of Liu and Cohen reads therefore as follows: *If* the material β-C_3N_4 exists, *then it should be superhard.* It should, nevertheless, be pointed out that the prediction is not limited to the specific modification β-C_3N_4; it also holds for any crystalline tetrahedrally bonded carbon nitride modification with the appropriate short bond length which might be realized. β-C_3N_4 has been chosen as a possible candidate for such a modification mostly by analogy with the well-known β-Si_3N_4; according to the calculations of Liu and Cohen it should be at least metastable.

During recent years, however, some doubt has been expressed repeatedly concerning the stability of β-C_3N_4 with respect to other possible tetrahedral crystalline CN modifications such as α-C_3N_4 or cubic C_3N_4 [27, 98]. Nevertheless, as this discussion has been carried out up to now only on the basis of theoretical calculations, and experimental data are not available yet, the term β-C_3N_4 will be used in the following as a synonym for any form of crystalline, diamond-like carbon nitride irrespective of its modification.

After the publication of the first paper of Liu and Cohen [22], intensive experiments have been carried out worldwide in an attempt to synthesize the material β-C_3N_4; in most cases, thin-film technology methods have been applied (Sect. 3.5). However, although the deposition of β-C_3N_4 has been claimed in a number of published papers, in almost all cases these claims do not withstand a thorough examination. At the same time, a review of the existing literature (Sect. 3.5) shows that, in the field of the deposition of *crystalline* carbon nitride films, two major problems exist: in almost all cases, the deposited $C_{1-x}N_x$ films are amorphous, containing at best some (nano)crystalline particles; furthermore, the maximum nitrogen content is in most cases between 30 and 40% which is considerably lower than the concentration of 57% required for β-C_3N_4.

This means that the state of the art of β-C_3N_4 deposition (as compared to the case of c-BN) is lagging behind by another important step: up to now it has not been possible to prepare this material at all (or at best only in the form of some isolated nanocrystals). The films obtained so far possess neither the crystallinity nor the stoichiometry required for β-C_3N_4.

An important problem in this context is that – in contrast to the cases of diamond (natural and synthetic) and c-BN (HPHT material) – there is no bulk material to compare the deposited films with. Many properties of the hypothetical compound β-C_3N_4 are still unknown. This implies difficulties concerning the characterization of the deposited carbon nitride films. At present there is no prescription for the identification of β-C_3N_4 of the kind that has been discussed above for the other superhard materials considered in this book. On the one hand, the required stoichiometry is known; in addition, the X-ray diffraction and electron diffraction patterns can be predicted from the calculated lattice constants. The comparison with the case of c-BN [72, 99] suggests, on the other hand, that these two properties are not sufficient for an identification beyond doubt. Additional characterization by techniques such as Raman, infrared or X-ray photoelectron spectroscopy is therefore necessary. Here we have the interesting case that the knowledge of the spectra of a material has to precede the preparation of this very material [100]; it is therefore necessary or at least desirable to calculate the bond energies and vibration spectra of a material theoretically without having at the present time the possibility to compare these calculations with experiment [101].

Nevertheless, it has to be pointed out here once again that β-C_3N_4 represents only a proposition of Liu and Cohen [22] for possible C_xN_y materials; it is therefore imaginable that in the C/N system other compounds with interesting, unusual properties exist. Indeed, some of the almost exclusively amorphous C_xN_y materials that have been realized possess properties [102, 103, 104, 105] calling for further investigation, although neither the stoichiometry nor the crystallinity of β-C_3N_4 have been obtained.

2.5 Organization of the Remainder of this Book

In the following two chapters, the deposition of the four diamond-like superhard materials considered in this book is discussed in detail. For this purpose it turns out to be necessary to distinguish between the formation of the first nuclei of the individual materials (nucleation, Chap. 4) and the growth after formation of these first nuclei (Chap. 3). The aim of the book is to provide a comprehensive presentation of this topic; nevertheless, emphasis is given to those aspects which have been investigated during recent years by the Thin Film Technology group at the University of Kassel:

- **Diamond**
    - Gas-phase processes during diamond deposition in the C/H/O system (Sect. 3.2.2).
    - Bias-enhanced nucleation of diamond (Sect. 4.3).
    - Applications of diamond films, e.g. as thin membranes [106] or as materials for sensors for scanning-probe microscopy applications [107, 108, 109].

- **Cubic Boron Nitride**
    - Growth of c-BN (Sect. 3.3).

⋄ Nucleation of c-BN (Sect. 4.4).
⋄ Modeling of growth and nucleation of c-BN (Sects. 3.3 and 4.4).
⋄ Investigations of actual problems of c-BN deposition (stress, adhesion, incorporation of hydrogen; Sect. 3.3).

- **Tetrahedral Amorphous Carbon**

    ⋄ Modeling of ta-C deposition (Sect. 3.4).

For those topics not covered by the work of our group, the following presentation relies on a summary of the existing literature. As a result, the selection is necessarily subjective. In addition, the situation of the research concerning the individual materials has to be taken into account. In the field of cubic boron nitride the number of groups, as well as the papers published up to now, are (still) limited; at the present time it is still possible to keep an overview over the published literature and the topics treated therein[19]. For diamond, on the other hand, this is by far not the case. Every year, some hundreds of papers on the low-pressure synthesis of diamond are published, covering topics ranging from the characterization of pure diamond surfaces, through experimental and theoretical investigations of the nucleation and growth of diamond, and the characterization of the deposited films with a wide variety of methods, to applications and the fabrication of the first devices.

This leads to the consequence that the presentation concerning the low-pressure synthesis of diamond is to a far greater extent based on the publications of other groups, and that it is by far more selective than is the case for cubic boron nitride.

---

[19] This also holds for ta-C and β-C_3N_4.

3. Growth Mechanisms

3.1 On the Deposition of Thin Films

The methods to deposit thin films from the gas phase[1] are principally divided into chemical vapor deposition (CVD) and physical vapor deposition (PVD) techniques. At the beginning of the development of thin-film technology, this separation was unequivocal (e.g. CVD: thermal decomposition, pyrolysis; PVD: evaporation); with the invention of plasma techniques, however, this sharp distinction could no longer by applied, since here chemical as well as physical processes play a role. Therefore, today, all those techniques where the source compounds are introduced as gases or vapors into the reaction chamber are labeled as CVD; in PVD methods, on the other hand, at least one of the source materials is introduced in solid form into the reactor, where it is transferred into the gas phase by physical means (e.g. evaporation or sputtering) [11, 112, 113][2]. Owing to distinct differences with respect to the expense of the apparatus, the working pressures, the possibility of three-dimensional deposition, the step coverages and the process costs, the distinction seems still to be justified, although the development of combination techniques has rendered the separation between PVD and CVD somewhat equivocal.

Independently of the type of method applied, for deposition techniques from the gas phase one can distinguish between gas phase processes, transport processes and surface processes. The source compounds are introduced into the gas phase of the reactor, where they react with each other or are decomposed. The products of these processes then reach the substrate, where they finally interact with the substrate surface. However, the importance of each of these steps varies depending on the method chosen. Finally, it should

---

[1] In addition, in thin-film technology methods exist for deposition from the liquid phase, such as LPE (liquid-phase epitaxy) [110], the sol–gel technique [111] and electrochemical (e.g. galvanic) techniques; however, these are not of importance in the context of this book.

[2] A further criterion to distinguish between CVD and PVD, based on thermodynamic considerations, has been proposed by Yarbrough [69].

be mentioned that the individual process steps can take place either spatially well separated or within a limited, small volume[3].

3.2 Growth of Diamond Films

3.2.1 Methods for Low-Pressure Diamond Synthesis

In 1953, diamond was, for the first time, synthesized successfully by means of a high-pressure/high-temperature method in accordance with the carbon phase diagram shown in Fig. 2.1. Some months before, however, Eversol deposited diamond from the gas phase by means of a low-pressure CVD technique. Further successful work on diamond CVD was carried out in the 1960s and 1970s in the US by Angus and coworkers and in the USSR by Deryagin's group. Nevertheless, the growth rates obtained during these years were extremely low; moreover, deposition was only possible on diamond substrates. Since 1970, on the other hand, a drastic increase of the deposition rates has been obtained; furthermore, Deryagin et al. were able to deposit on other substrate materials. Despite these successes, however, these investigations were hardly noted by the scientific community, until the Russian work was verified in Japan by a research group at NIRIM at the beginning of the 1980s[4], among others.

Ever since this verification of the Russian experiments in Japan, the interest in low-pressure synthesis of diamond has risen almost exponentially. Likewise, a vast variety of methods for the deposition of diamond have been developed which – apart from the fact that without exception CVD techniques are applied – rely in first instance on very different principles. The most important of these techniques are summarized in Table 3.1; a more detailed discussion can be found in [85].

Despite all differences, the methods listed in Table 3.1 have some important features in common:

1. All methods of diamond deposition are marked by a high gas phase temperature T_G[5] and a distinctively lower substrate temperature T_s which

---

[3] One method in which these three steps are spatially clearly separated from each other, and by which they can therefore be characterized independently of each other, is the remote PECVD (R–PECVD) method, which has been studied intensively by the author's group [114, 115].

[4] An overview of these early investigations and the appropriate references can be found in [84, 116, 117].

[5] In the first instance, the definition of the gas phase temperature T_G is not without problems. For some methods, such as the hot-filament technique ($T_G = T_{\text{filament}}$) or oxyacetylene flames, it can be given directly. For plasma methods such as MWCVD, however, the temperatures of the various ensembles (ions, electrons, neutrals) are different. For these techniques, the definition of T_G follows from the discussion in Sect. 3.2.2.

Table 3.1. Compilation of typical process parameters (pressure p, gas phase temperature T_G and power used P) for the most important techniques of low-pressure diamond synthesis. These data, as well as the area coated A and typical growth rates R, have been taken from [85]. They represent average values of the state of the art in ca. 1991; since then, most techniques have been improved with respect to the coatable area and the growth rates obtained. The last two columns list the mass deposition rate R' in units of carat/h and the energy consumption for the deposition of one carat of diamond, P/R'. These values have remained almost unchanged despite all new developments since 1991. MW, micro wave

Method	Parameters			Deposition			
	p (mbar)	T_G (°C)	P (kW)	A (cm^2)	R (μm/h)	R' (c/h)	P/R' (kWh/c)
RF plasma	0.1–40	1000–1500	0.5–3	300	0.1	0.02	50
Hot filament	20–80	2000–2400	0.1–1	1–5	1–10	0.035	15
MW plasma	20–100	2000–2500	0.5–2	30	1–5	0.03	30
Arc/jet	1–30	> 4000	5–20	2	25	0.08	80
Oxyacetylene flame	10^3	3100	1	0.5	50–100	0.09	10
Thermal MW plasma	10^3		5	5	10–30	0.1	50
Thermal RF plasma	10^3	> 4000	6–60	3	120–180	1	30
Arc/jet	10^3	> 4000	1000	90	500–1000	150	10

is decoupled from T_G [85]. The individual techniques are distinguished basically by the method used to provide gas phase temperatures as high as possible (filament, various plasmas and flames).

2. The substrate temperatures are, for all methods, typically between 700 and 1000°C[6]. Working pressures are either in the range of 10 to 100 mbar or at atmospheric pressure. Typical process gases consist of highly diluted hydrocarbons in hydrogen (e.g. CH_4, C_2H_2; the carbon content of the gas phase is 0.1–5%). The gas flows scale roughly with the mass deposition rates listed in Table 3.1[7].
3. The growth rates are higher the higher the gas phase temperature (Fig. 3.1).
4. The energy required for the deposition of one carat of diamond is almost independent of the method applied (Table 3.1, last column); it lies between 10 and 100 kWh/c[8].

These common features, especially the almost constant energy input per carat of diamond deposited, suggest that, despite all differences in apparatus

---
[6] Exceptions will be discussed and explained below; they are, in addition, independent of the method applied.
[7] MW plasmas, hot filament: 100–500 sccm; arc plasmas at 1 bar: 100 000 sccm!
[8] Simultaneously, the power used varies by about four orders of magnitude!

46   3. Growth Mechanisms

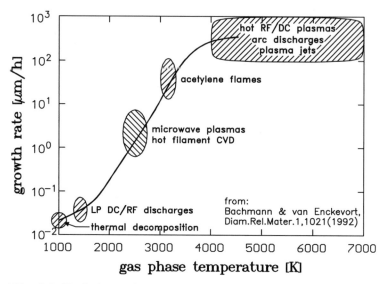

Fig. 3.1. Typical growth rates and gas phase temperatures for the most important methods of diamond synthesis (according to [85])

and parameters, all methods are based on some common basic mechanisms. The influence of T_G on the growth rate shows, further, that gas phase processes play a decisive role in diamond deposition. Finally, it is evident that the coatable area depends on the capability of each method to provide a high gas phase temperature over a large area.

Table 3.2. (Possible) roles of atomic hydrogen during the low-pressure synthesis of diamond. For most aspects, a series of sources can be found, e.g. the overview articles [54, 55, 83, 88]. The emphasized point 'regulation of gas phase composition' was — at least in this formulation — developed by the author's group; it is discussed below in Sect. 3.2.2 in detail

Gas phase processes	Transport processes
⋄ Creation of active carbon species ⋄ **Regulation of gas phase composition**	⋄ Prolongation of the lifetime of active carbon species ⋄ Transport of energy

Growth	⟵ Surface processes ⟶	Nucleation
⋄ Stabilization of diamond surfaces ⋄ Creation of growth sites on the surface ⋄ Etching of sp^2 carbon ⋄ Hydrogen abstraction from carbon species on the surface		⋄ Stabilization of diamond nuclei ⋄ Reduction of the critical nucleus size ⋄ Suppresion of sp^2 nuclei ⋄ Hydrogen termination of the substrate

The present models of low-pressure diamond synthesis assume without exception that high gas phase temperatures are necessary to provide a partial pressure of atomic hydrogen as high as possible. Likewise, it is generally assumed that atomic hydrogen has to fulfill not one but several roles during diamond deposition. The most important are summarized in Table 3.2; they will be discussed in detail in the following sections.

3.2.2 Gas Phase Processes

In the following, the term 'gas phase processes' will include all phenomena taking place in the region of high gas phase temperatures mentioned above. In the case of the plasma processes this region can be defined spatially rather easily; its limits are given by the spatial extent of the plasma. In other cases (e.g. for hot-filament CVD, HFCVD) this definition is somewhat more difficult since the region of high temperatures is relatively small and not sharply separated from the rest of the gas volume.

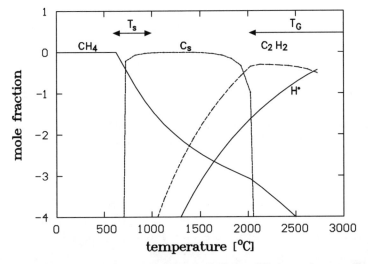

Fig. 3.2. Temperature dependence of the equilibrium concentrations of the most important species during diamond deposition for 0.1 mol CH_4 in 10 mol H_2 at 40 mbar. The carbon-containing species have been normalized with respect to the total carbon concentration, and [H$^\bullet$] with respect to the total hydrogen concentration. The *arrows* indicate typical regions of the substrate temperature (600–1000°C) and the gas phase temperature ($\geq$ 2000°C). C_s represents solid carbon

Thermodynamic Aspects. A preliminary description of diamond growth can be carried out using equilibrium considerations following Anthony [54] and Sommer and Smith [118, 119]. In Fig. 3.2 the concentrations of the most important carbon species (for 1% CH_4 in H_2) are shown as a function of temperature for a heterogeneous equilibrium system. At low temperatures (< 1000°C) the gas phase consists almost exclusively of CH_4. With increasing temperature, more and more methane is decomposed exothermically by reaction with hydrogen, thereby creating condensable species. Simultaneously, the 'solubility' of carbon in the hydrogen gas phase decreases [54] (in other words an increasing amount of carbon is present in solid form). If the temperature is increased further, endothermic formation of C_2H_2 sets on; the solubility increases again and, consequently, the portion of carbon present in solid form decreases. The typical substrate temperatures during diamond deposition exactly match this minimum of carbon solubility. This explains, at least in terms of tendencies, the observed growth rates [54]: the deposition takes place at the heated substrate and not on the considerably cooler walls of the reactor, where the gas phase solubility is higher. Likewise, the differences between the growth rates for techniques with different gas phase temperatures (RF plasmas, hot filament, thermal plasmas; see Fig. 3.1) can be explained by this temperature dependence of the solubility of carbon in hydrogen. Taking the carbon alone, according to this argument diamond synthesis would also be possible for conditions where $T_G < T_s$; however, a separate source of atomic hydrogen would be required in order guarantee that the deposited carbon consists indeed of diamond. The equilibrium concentration of $H^{\bullet}$ is also shown in Fig. 3.2.

These considerations have been extended by Sommer and Smith [118, 119]. They also start from an equilibrium between a solid carbon surface and gaseous species desorbing from this surface. In contrast to the calculations presented above, however, they distinguish explicitly between diamond and graphite. Furthermore, they do not consider concentrations but, rather, desorption rates for gaseous species and adsorption rates for solid modifications:

$$R^{\text{ads}}_{C_xH_y} = \frac{\eta P_{C_xH_y}}{2\pi m k_B T} \quad \text{and} \quad R^{\text{des}}_{C_xH_y} = \frac{\eta' P^{\text{eq}}_{C_xH_y}}{2\pi m k_B T}, \qquad (3.1)$$

where η and η' are (non-thermodynamic) sticking and desorption coefficients, respectively; P is the partial pressure of impinging C_xH_y species and P^{eq} the corresponding equilibrium partial pressure. From this, the net carbon deposition rate

$$R_C(T_s, p, C_c) = \sum_{x,y} R^{\text{ads}}_{C_xH_y} - \sum_{x,y} R^{\text{des}}_{C_xH_y} \qquad (3.2)$$

can be determined ($C_c = [C]/([C]+[H])$ denotes the carbon content of the gas phase). The stability curve $R_C(T_s, p, C_c) = 0$ then separates the parameter regimes where only gaseous species can exist from those regions where a solid carbon phase is present.

The results of these calculations are presented in Fig. 3.3, showing stability curves for graphite (curve 1) and diamond (curve 2) as a function of temperature and the carbon content of the gas phase, thereby assuming all sticking and desorption coefficients to be unity. On the right-hand side of

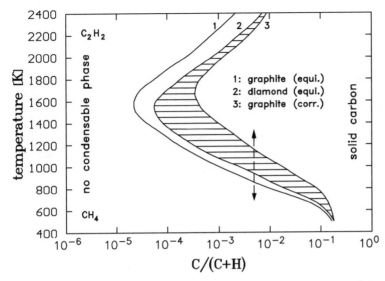

Fig. 3.3. Model of Sommer and Smith: stability limits of graphite (1) and diamond (2) under equilibrium conditions, and for graphite after correction for the desorption coefficients (3) [119] (at 50 mbar). The standard concentration of 1 % CH_4 in H_2 is indicated by the *vertical dashed line*

the curves solid carbon is stable; on the left-hand side, only gaseous species exist. Independent of temperature, the stability region of diamond is always confined within that of graphite, which is a consequence of the metastable nature of diamond. According to these calculations, therefore, deposition should always yield graphite. In addition, the temperature dependence of these stability curves corresponds to that of the solubility curve in Fig. 3.2.

This means that thermodynamic considerations alone are not sufficient to explain diamond growth. Rather, kinetic aspects also have to be taken into account. To this end, Sommer and Smith change somewhat arbitrarily the desorption coefficient η'_{H_2}(gra) from 1 to 0.2. This corresponds to an enhanced etching of graphite by atomic hydrogen, thereby forming CH_4, which is indeed observed experimentally. The resulting stability curve for graphite is also shown in Fig. 3.3 (curve 3). It is now completely confined within the stability curve for diamond; thus a region exists where the deposition of diamond is possible (hatched in Fig. 3.3).

With this modification, the model of Sommer and Smith predicts a number of experimental observations correctly:

- For 1 % CH_4 in H_2, a temperature window for diamond deposition from approximately 700 to 1000°C (dashed line in Fig. 3.3) results, which is in agreement with the experimental facts.
- Towards higher temperatures (approximately 1000°C) a transition to graphite takes place, which is indeed obeserved (Sect. 3.2.4).

- Likewise, an increase of the methane concentration at constant temperature results in a shift towards graphite. This is again in agreement with experiment.
- In addition, a second window for diamond deposition exists in the high-temperature region ($< 2000°C$). This window is indeed observed experimentally; as for too low a filament temperature, a deposition takes place on the filament itself (see e.g. [106]).

Thermodynamics versus Kinetics. The arbitrary variation of the desorption coefficient in the model of Sommer and Smith indicates that thermodynamic considerations alone cannot explain the growth of diamond; nevertheless, it has to be noted that these considerations already allow a rather deep insight into the mechanisms of diamond deposition.

Especially at the beginning of the 1990s, there has been an intensive discussion as to whether the low-pressure synthesis of diamond is dominated by thermodynamic or kinetic processes (see e.g. [52, 55, 83, 88, 120] and especially the work of Yarbrough [69, 86, 121, 122]). A detailed presentation of this discussion is beyond the scope of this book; generally, it is assumed today that diamond deposition is driven kinetically. However, in this context it has to be taken into account that all thermodynamic discussions, especially those concerning stability questions, have been almost exclusively limited to the carbon system. In fact, the carbon/(atomic) hydrogen system has to be considered; for this system, the data for gas phase species are wellknown, but not those for surface species and, most importantly, for hydrogen-stabilized surfaces. Badziag et al. [123], for example, have shown that hydrogenated diamond particles with diameters of some nanometers are more stable than the corresponding sp^2-bonded clusters. A final resolution of the question of thermodynamics versus kinetics seems, therefore, to be possible only when sufficient thermodynamic data for the carbon/atomic hydrogen system become available.

Reaction Mechanisms. Most techniques for diamond deposition work with a mixture of a hydrocarbon (in most cases CH_4) and hydrogen; typically, the carbon content of the gas phase is between 0.1 and 5 %. In the first step, hydrogen is dissociated in the plasma or at the filament:

$$H_2 \xrightarrow{\text{heat,plasma}} 2H^\bullet . \tag{3.3}$$

The atomic hydrogen $H^\bullet$ thus generated reacts with the hydrocarbons introduced into the gas phase (e.g. CH_4), thereby creating active, condensable species (e.g. $CH_3^\bullet$, C_2H_2) via reactions of the kind

$$CH_4 + H^\bullet \longrightarrow CH_3^\bullet + H_2 , \tag{3.4}$$
$$CH_3^\bullet + CH_3^\bullet + M \longrightarrow C_2H_6 + M , \tag{3.5}$$
$$C_2H_x + H^\bullet \longrightarrow C_2H_{x-1} + H_2 \tag{3.6}$$

(where M is a third body). All of these reactions with $H^\bullet$ are strongly exothermic. The recombination of atomic hydrogen

$$H^\bullet + H^\bullet + M \longrightarrow H_2 + M \,, \tag{3.7}$$

on the other hand, requires a third body and is thus rather improbable.

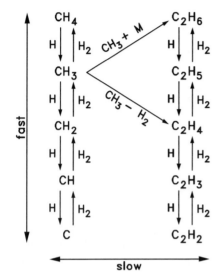

Fig. 3.4. Schematic illustration of the gas phase reactions during diamond deposition in the C/H system (according to [88])

Figure 3.4 shows a possible reaction scheme based on the reactions (3.3)–(3.6) [88] (in the figure, species with more than two carbon atoms have been neglected). For the discussion in Sect. 3.2.2 it is of great importance that the mechanism presented in Fig. 3.4 is based exclusively on chemical reactions. Plasma-induced processes such as electron-impact dissociation of carbon-containing species do not play a role.

The H/C/O System. The working gas of diamond deposition consists — as outlined above — usually of a highly diluted hydrocarbon/hydrogen mixture. A vast variety of investigations (see e.g. the references listed in [124]) shows, on the other hand, that diamond deposition is also possible with a great number of oxygen-containing C/H/O mixtures. In addition, it has been demonstrated in recent times that hydrogen can be replaced partly or even completely by N_2 [125, 126] or halogens [127, 128, 129]; nevertheless, in the following, only the H/C/O system is considered in order to keep the presentation short.

An analysis by Bachmann et al. [124] of the data accumulated before 1991[9] shows that the C/H/O system can be divided into three regions (Fig. 3.5). Diamond growth is only possible within a narrow region (the diamond

---

[9] Later investigations have confirmed, without any exceptions, the conclusions drawn by Bachmann et al.

52   3. Growth Mechanisms

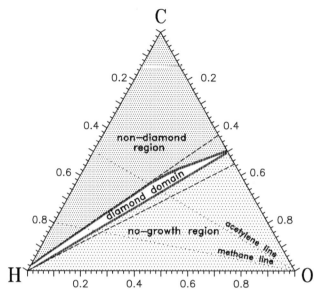

Fig. 3.5. Growth regions in the C/H/O system (Bachmann diagram). The original data collection (*dashed lines*) was published in [124]. The revised limits of the diamond domain were obtained from [130]

domain) above the line C/O = 1 (CO line)[10]. If the carbon content is too high, non-diamond-bonded carbon is obtained; for too high oxygen contents, on the other hand, no growth of any carbon film takes place. The nature of the source gases does not play a role [106]; the domains are determined exclusively by the C/H/O ratio.

The Bachmann diagram represents, nevertheless, a compilation of data only; it provides no explanations of the mechanisms it is based on. The author' group has tried to evaluate these mechanisms by intensive deposition experiments with HFCVD and MWCVD [131, 132, 133], gas phase investigations with optical emission spectroscopy (OES) and mass spectrometry (MS) [134, 135], and equilibrium calculations of the gas phase composition [106]. This work leads to a model of the gas phase processes during diamond deposition in the C/H/O system [106, 136, 137], which is presented briefly in the following. In this model, only the region of the gas phase is considered which is marked by high gas phase temperatures T_G. For MWCVD this concerns the extent of the plasma ball, for HFCVD the direct vicinity of the filament. The transport of species from the hot gas phase to the considerably cooler substrate is not taken into account at this point of the discussion.

We will start by listing some general observations and correlations which we obtained from our investigations of the C/H/O system [106, 134, 135, 136]:

---

[10] The shape of the domain originally postulated in [124] has been revised in later publications (e.g. [130]). This has been taken into account in Fig. 3.5.

1. The concentration of atomic hydrogen [H] is roughly constant within the diamond domain and in its direct vicinity. This is also evident from the equilibrium calculations presented in Figs. 3.6–3.9. The only exception is the hydrogen-poor region in the vicinity of the C–O baseline. This means that [H] cannot be responsible for the existence, localization and shape of the diamond domain. The nature and properties of the deposited films are determined, rather, by the concentrations of the various hydrocarbon species and their ratios among each other and to [H].
2. The relative concentrations of the monomeric carbon species (CH, CH_3, CH_4) correlate, in wide ranges of the C/H/O system, with each other as well as with the growth rates. The relative concentrations of the dimeric carbon species (C_2, C_2H_2), on the other hand, correlate inversely with the diamond quality.
3. The equilibrium calculations are in good agreement with the parameter dependences observed during gas phase investigations by means of OES and MS[11]. Therefore, the following discussion can be carried out on the basis of these calculations. It should, nevertheless, be pointed out that all dependences of the gas phase composition in the C/H/O system presented in the following have also been observed experimentally [106, 134, 135, 136].

Figure 3.6 shows equilibrium calculations alongside the C–H baseline in the vicinity of the hydrogen corner (Fig. 3.5)[12]. On reduction of the carbon content of the gas phase, the concentrations of all carbon-containing species decrease, as is to be expected. Nevertheless, the decrease of the acetylene concentration is much more pronounced than that of the monomeric species CH_3 and CH_4. This becomes especially evident from the normalized presentation on the right-hand side of Fig. 3.6. Taking into account the fact that – as described above – the concentrations of monomeric carbon species correlate with the growth rate, and those of dimeric species, on the other hand, inversely with film quality, a 'diamond window' (i.e. a section through the diamond domain) becomes apparent in Fig. 3.6 where diamond growth is possible with, however, increasing degradation of the film quality.

The results presented in Fig. 3.6 concern the oxygen-free case, i.e. the standard mixture for diamond deposition. In order to determine the influence of oxygen, several section through the diamond domain normal to the carbon monoxide line (CO line) have been investigated (Fig. 3.5). These yield very similar results, independent of the individual H/C values. Figures 3.7 and 3.8

---

[11] Under the conditions of diamond deposition, OES and MS allow only relative measurements; an absolute determination of the gas phase composition is not possible with these methods.

[12] These and all following calculations were performed for a pressure of 26 mbar and a gas phase temperature of 2500 K. The pressure corresponds to that of the experiments carried out in parallel; the choice of T_G will be justified below.

54   3. Growth Mechanisms

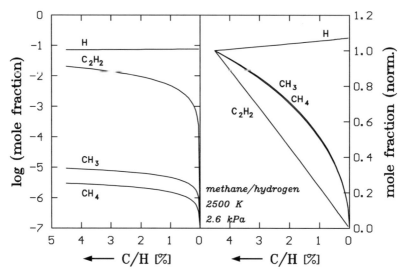

Fig. 3.6. Concentration of the relevant species along the C–H baseline. *Left*: logarithmic presentation of the absolute values. *Right*: normalized linear presentation of the same data. Note that the x axis is plotted from *right* to *left* in order to facilitate comparison with Figs. 3.7 and 3.8

present as examples calculations for H/C = 84 and H/C = 0.43, respectively, i.e. for the hydrogen-rich and hydrogen-poor regions of the C/H/O system.

On approaching the CO line, in each case the same behavior is observed as for the reduction of the carbon concentration in the oxygen-free case presented in Fig. 3.6: The concentrations of all hydrocarbon species decrease drastically, but again, the decrease of the concentration of dimeric species is much more pronounced[13]. Normalized linear presentations of these data sets would therefore show the same 'diamond window' as in Fig. 3.6. From Figs. 3.7 and 3.8 it is further evident that surplus carbon is bonded as carbon monoxide. CO is the most stable diatomic molecule; its dissociation, as well as reactions releasing the carbon atom, is thus rather improbable. It is therefore realistic to assume that carbon atoms bonded as CO do not contribute to diamond growth.

Figure 3.9, finally, shows a section alongside the CO line through the entire domain (cf. Fig. 3.5). It can be seen that the concentrations of the most important species vary, at least up to C/(C+H) $\approx$ 0.75, by one order of magnitude at the most. The only exception is CO, which again serves as a drain for surplus carbon.

At this point it should be pointed out once again that all the important results which have been derived from the equilibrium calculations presented in Figs. 3.6–3.9 ([H] is roughly constant within the domain; a 'diamond window'

---

[13] This behavior was also observed by Prijaya et al. [138].

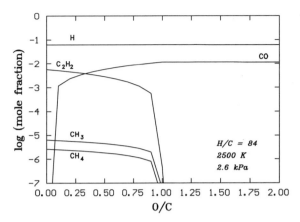

Fig. 3.7. Section through the domain in the hydrogen-rich corner of the C/H/O system (H/C = 84, i.e. 2 % CH$_4$/H$_2$). Presentation of the dependence of the most important species on the O/C ratio

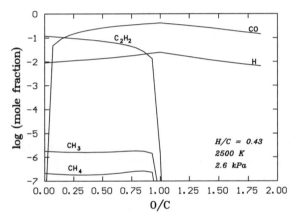

Fig. 3.8. Section through the domain in the hydrogen-poor region of the C/H/O system (H/C = 0.43). Presentation of the dependence of the most important species on the O/C ratio

exists for the oxygen-free case as well as for the various sections through the domain) have been confirmed experimentally, by means of gas phase diagnostic measurements in MWCVD and HFCVD reactors by the author's group as well as by other authors [139, 140, 141].

Discussion. The equilibrium calculations presented above and their experimental verification allow some very important conclusions concerning the role of oxygen in the gas phase during diamond deposition, the existence of the diamond domain in the Bachmann diagram (Fig. 3.5), and the low-pressure synthesis of diamond in general.

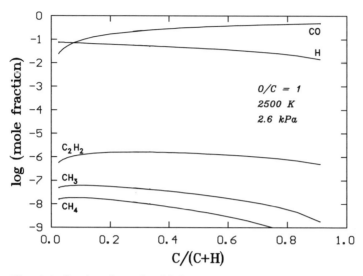

Fig. 3.9. Section along the CO line

Role of Oxygen. First of all, the most important role of oxygen in the gas phase consists in the regulation of the concentration of active carbon species via reactions of the type

$$C_{condensable} + O \longrightarrow CO . \tag{3.8}$$

This means that the concentrations, or rather the ratios of the concentrations, of the relevant species ($H^\bullet$, C_1H_x, C_2H_y) are almost constant within the domain along lines which are roughly parallel to the CO line (Figs. 3.6–3.8). In other words, reactions of the type (3.8) project the diamond domain on the C–H baseline[14]. Simultaneously they explain the 'no growth' region for $[O]/[C] > 1$.

The conditions for the oxygen-free deposition of diamond and for deposition from oxygen-containing gas phases are therefore identical if one considers only the concentration of *free* carbon

$$[C]_f = [C] - [O] \tag{3.9}$$

instead of the total concentration of carbon in the gas phase. This concept of considering the free carbon only has now been confirmed by other groups (e.g. [88, 138, 142, 143]) and is now generally accepted.

The most important role of oxygen in the gas phase consists therefore in the regulation of the concentrations of condensable carbon species. If this is the task oxygen species perform during diamond deposition, neither the growth rate nor the film properties should be influenced by the addition of

---

[14] The deviations in the hydrogen-poor region in the vicinity of the C–O baseline, mentioned above, explain in addition the narrowing of the diamond domain in this part of the Bachmann diagram (Fig. 3.5).

oxygen; both should be determined exclusively by the concentration of free carbon $[C]_f$. This, however, is not the case:

- The addition of oxygen increases the growth rate, at least in the hydrogen-poor regions of the diamond domain. However, this effect is not nearly as pronounced as postulated in some early publications on the addition of oxygen [144, 145, 146]. Our own investigations [106] show, for example, that alongside the CO line the rates are constant within $\pm 10\%$ from H/C = 44 to H/C = 4; deviations were observed only for very low values of H/C.[15]
- The addition of oxygen to the gas phase shifts the low-temperature limit of diamond growth to considerably lower substrate temperatures with respect to the oxygen free case (while the diamond quality remains high) [145, 146, 147].
- Our own investigations [136] show, furthermore, that the addition of oxygen has a pronounced effect on the texture, morphology and crystallite size of the deposited films, in good agreement with the results of other groups [142, 147, 148].

From these observations, it must be concluded that oxygen species must also play a role during surface processes. This aspect will be addressed later in Sect. 3.2.4.

Existence of the Diamond Domain. The results presented above and the role of oxygen during gas phase processes derived from these results further suggest that the existence of the diamond domain in the Bachmann diagram is a result of thermodynamic equilibrium processes. It is obvious, however, that the 'equilibrium temperature' has to be much higher than the substrate temperatures typically used (cf. Fig. 3.2).

Bou et al. [149] have shown by means of kinetic calculations that in microwave plasmas under conditions of diamond deposition the dissociation of H_2 is caused a large extent by electron impact processes; the hydrocarbons, on the other hand, are dissociated mainly by reactions with atomic hydrogen (see reaction (3.4) and the reaction scheme in Fig. 3.4). Thus the plasma can be regarded essentially as a source of atomic hydrogen (compare [55, 88, 150]); on the other hand, the gas phase chemistry and, as a consequence, the gas phase composition are determined by reactions with atomic hydrogen (see also [88, 151]).

Thus, according to our model, *the gas phase composition of an MW plasma under conditions of diamond deposition corresponds to the equilibrium composition at a temperature of* $T_G = T_{H^\bullet}$, *which is determined by*

---

[15] A critical retrospective examination reveals that the drastically enhanced growth rates reported in the early investigations, which were to a great extent responsible for the recent intensive exploration of the C/H/O system, can be explained to a large extent by the fact that, in addition to the use of oxygen, further parameters were changed [144, 145, 146].

the concentration of atomic hydrogen $H^{\bullet}$, which in turn is determined by the plasma conditions[16]. The same considerations are also valid for hot-filament deposition. Here, the filament acts as the source of atomic hydrogen; its conditions determine [H] and $T_{H^{\bullet}}$. These considerations relate, of course, not only to the deposition of diamond with the addition of oxygen but also to the oxygen-free case which – although much more common – presents nothing but a special case of the H/C/O system.

Since, according to the above findings, the Bachmann diagram reflects the fact that the gas phase processes can be described by a thermodynamic equilibrium, deviations from this diagram should occur particularly in those cases where the equilibrium conditions are not fulfilled. For plasma processes, this concerns especially processes taking place at low pressures and/or low input powers. It can indeed be shown that under such conditions diamond deposition is possible outside the diamond domain given by the Bachmann diagram:

1. For microwave deposition, for a given C/H/O ratio the gas phase composition and the growth rate depend on the nature of the source gases if low powers and/or low pressures are used [134].
2. For hot-filament CVD, the deposition of high-quality diamond is possible outside the diamond domain (within the nominal no-growth region) if very low filament–substrate distances (≤ 0.5 mm) are used [106, 132].
3. Further examples can be found in the literature; a thorough analysis [152] shows that at very low pressures diamond deposition is indeed possible within the nominal 'no growth' region of the Bachmann diagram (e.g. [153, 154, 155]).

It should, finally, be mentioned that Prijaya et al. [138] (and also Benndorf et al. [139]) present equilibrium calculations very similar to those described above; furthermore, these authors reach the same conclusions [138, 156, 157] concerning the role of oxygen in the gas phase and the importance of equilibrium thermodynamics for the existence of the diamond domain within the Bachmann diagram.

Conclusions. The preceding paragraphs have shown that the gas phase processes during low-pressure synthesis of diamond are understood to a large extent. The plasma or filament acts as a source for the creation of atomic hydrogen; the concentration [H] in turn regulates the concentrations of the other species (it has to be kept in mind that in the case of the addition of oxygen, only the concentration of free carbon is of relevance). The gas phase can therefore be described to a good approximation by an equilibrium system; nevertheless, it has to be pointed out again that this equilibrium corresponds to a fictive temperature $T_{H^{\bullet}}$.

---

[16] This is also the basis for the choice of $T_G = 2500$ K for the above equilibrium calculations, as the concentration of atomic hydrogen in MW diamond plasmas is on the order of 5–10 % (compare with Fig. 3.2).

These facts also explain why the simple thermodynamic models by Anthony [54] and Sommer and Smith [118, 119] at the beginning of this section are able to deliver a relatively good description of diamond growth.

At the end of this section, it should be noted once again that the results and discussions presented above concern the gas phase exclusively, i.e. the region of high temperature T_G. It is of course obvious that the gas phase composition changes during transport from the hot gas phase to the cooler substrate. Here, the non-equilibrium character of diamond deposition becomes manifest. This aspect will be discussed in the following section. Nevertheless it is to be expected that at least some information about the original composition is maintained during the transport processes, as was also pointed out by Prijaya et al. [138]. In principle, the very existence of the diamond domain in the Bachmann diagram confirms this assumption.

3.2.3 Transport Processes

During low-pressure synthesis of diamond, excited species ($H^\bullet$, $CH_3^\bullet$) are created in the hot gas phase (filament or plasma), for which the cooler substrate represents a drain. As a consequence, in every case a diffusion layer (chemical boundary layer) is formed between the gas phase and substrate [88]. In addition, a momentum boundary layer is formed in front of the substrate for those methods working with high gas velocities (see below) [55, 88]; through this boundary layer, mass transport is again dominated by diffusion processes[17]. Transport processes therefore play an important role in low-pressure diamond synthesis, which is nevertheless often underestimated in the literature (but see e.g. [55, 88, 158, 159, 160]).

Diffusion and Convection. A preliminary insight into the underlying processes, as well as a first explanation of the observed differences in the growth rates of the various techniques of diamond deposition, can be obtained by some simple considerations of the nature of mass transport. The Peclet number

$$\text{Pe}_{\text{mass}} = \bar{v}L/D \tag{3.10}$$

is a measure of the relative importance of diffusion and convection for mass transport within a reactor; for low Peclet numbers diffusion dominates, whereas for high Pe, on the other hand, the mass transport is governed by convection ($\bar{v}$ is the average convective velocity, L a characteristic length and D the diffusion constant).

---

[17] For plasma processes such as MWCVD, in principle, the transport of species through the plasma sheath (space charge layer) has also to be taken into account. However, diamond deposition is dominated by radical neutral species, whereas charged particles are of minor importance at best; therefore, space charge effects can be neglected during diamond growth. This is, however, not true for the bias–enhanced nucleation of diamond discussed in detail in Sect. 4.3.

60   3. Growth Mechanisms

Table 3.3. Peclet numbers for mass transport for some important methods of diamond deposition. Also listed are the time constants for the mass transport τ, as well as typical growth rates $\bar{R}$. The typical length was in all cases assumed to 1 cm (e.g. filament–substrate or outlet nozzle–substrate distance)

Method	$\bar{v}$ (cm/s)	D (cm²/s)	Pe	τ (s)	$\bar{R}$ (μm/h)	Sources
HFCVD	1	500–1000	1×10^{-3}	2×10^{-3}	1	[55, 88, 152] [158, 161]
MWCVD			5×10^{-4}		1	[55]
Thermal RF plasma	10^3	10	100	2×10^{-3}	10	[158]
Plasma jet	10^4		300		100	[55]

Table 3.3 summarizes typical values of Peclet numbers for some diamond reactors. From the table it is evident that those methods working at pressures between 10 and 100 mbar (e.g. HFCVD and MWCVD, compare Table 3.1) are marked by low Peclet numbers; for these techniques the mass transport is dominated by diffusion. Those methods working at atmospheric pressure, simultaneously utilizing high gas velocities, on the other hand, possess high values of Pe; here the mass transport thus takes place to a large extent by (forced) convection.

For the diffusion-dominated methods (HFCVD, MWCVD) the time constant for the mass transport can be expressed as

$$\tau_D = L^2/D . \tag{3.11}$$

As shown in Table 3.3, τ_D is on the order of some milliseconds; mass transport from the gas phase to the substrate takes thus place very rapidly.

For those methods working at atmospheric pressure and with high gas velocities, the time constant of the transport consists of two contributions. First, forced convection takes place (τ_K) until the boundary layer is reached, through which the transport is again dominated by diffusion (τ_B). With the values in Table 3.3 [158], it follows for an atmospheric pressure RF plasma process that

$$\tau_K = L/\bar{v} \approx 1 \times 10^{-3} \text{ s} \quad \text{and} \quad \tau_B = d^2/D \approx 1 \times 10^{-3} \text{ s}, \tag{3.12}$$

where d is the thickness of the boundary layer. The time required for transport from the gas phase to the surface is therefore very similar for both types of process, although the transport mechanisms are quite different. Transport processes cannot therefore be responsible for the growth rates of the atmospheric plasma process, which are higher by one order of magnitude (Table 3.3). The reason for these higher rates can be found, rather, in the process pressure, which is higher by approximately one order of magnitude, and, in turn, the higher absolute concentrations of reactive species (C_xH_y, H•); in addition, the higher gas phase temperature may also play a role. This leads to a higher flux of growth species to the surface, while the transport time remains roughly the same.

Switching from the atmospheric-pressure RF plasma to the plasma jet brings about a further increase of the growth rate by another order of magnitude (Table 3.3). Since RF plasmas and the plasma jet both work at atmospheric pressure, the pressure argument no longer holds. Here the higher gas velocities (10^4 cm/s) come into play; τ_K and τ_B are inversely proportional to $\bar{v}$, as the thickness of the boundary layer d decreases as $\bar{v}^{-1/2}$. The differences between the rates of the atmospheric RF plasma and the plasma jet are therefore caused by the transport process.

These simple considerations concerning the mass transport thus explain the rates observed for the different methods at least qualitatively. In addition, they allow another important conclusion: although the transport mechanisms for methods such as MWCVD and HFCVD (diffusion), on the one hand, and the atmospheric-pressure techniques (convection), on the other hand, are totally different, there are no differences with respect to the product (well-faceted diamond layers) or the energy input required for the deposition of one carat of diamond (Table 3.1). It is therefore obvious that transport processes indeed influence the growth rates by providing reactive species, but that the essential growth mechanisms are independent of the nature of the mass transport. These mechanisms will be discussed in Sect. 3.2.4.

Transport of Atomic Hydrogen. It will be shown in Sect. 3.2.4 that the growth rates and qualities of diamond layers are determined by the concentrations of atomic hydrogen and methyl radicals at the surface. These concentrations, however, cannot be directly adjusted from outside; rather, they are coupled to the controllable parameters such as the gas flows, gas phase temperature and pressure via the transport processes [159]. For a quantitative description of diamond growth, especially for process optimization and scaling up, therefore, a knowledge of the transport of active species to the substrate is a sine qua non. Since [CH$_3$] is strongly coupled to [H] via reaction (3.4) [159], the transport of atomic hydrogen from the gas phase to the substrate is of special interest. This transport of atomic hydrogen has been discussed in great detail by Goodwin [159]; some essential aspects are briefly presented in the following.

Starting from some simple diffusion–theoretical considerations, Goodwin derives, for the ratio of the mole fractions of atomic hydrogen at the surface and at some remote reference point (filament surface, plasma bulk), the following expression:

$$\frac{X_H^s}{X_H^{ref}} = \left(1 + \frac{\gamma_H}{Kn}\right)^{-1} . \tag{3.13}$$

Here γ_H is the surface recombination coefficient (the probability that a hydrogen atom impinging on the surface will recombine); at 1200 K, γ_H is approximately 0.1 [162]. $Kn = \lambda_H/l_d$ is the Knudsen number, for which the characteristic diffusion length l_d is defined as follows:

62   3. Growth Mechanisms

$$\left(\frac{dX_H}{dz}\right)_{z=0} = \frac{X_H^{ref} - X_H^s}{l_d}. \tag{3.14}$$

The quantity $\lambda_H = 4D_H/\bar{v}_H$ is the mean free path for hydrogen transport (here, D_H and $\bar{v}_H$ are the diffusion coefficient and the average thermal velocity, respectively, of atomic hydrogen); λ_H depends on the pressure, temperature and gas phase composition but is independent of the gas velocity.

If $Kn \gg \gamma_H$, it follows from (3.13) that $X_H^s = X_H^{ref}$, i.e. there is no gradient of the hydrogen mole fraction. In the opposite case ($Kn \ll \gamma_H$), the hydrogen mole fraction at the substrate surface is much smaller than at the point of reference:

$$\frac{X_H^s}{X_H^{ref}} = \frac{Kn}{\gamma_H} = \frac{\lambda_H}{l_d \gamma_H}. \tag{3.15}$$

Under such conditions, the hydrogen transport is limited by diffusion. In the following, we have therefore to distinguish between diffusion-dominated (HFCVD, MWCVD) and convection-dominated (oxyacetylene flame, plasma jet) methods.

Table 3.4. Typical parameters of a hot-filament reactor.

D_H (cm/s^2)	$\bar{v}_H$ (cm/s)	r_f (cm)	d_f (cm)	γ_H	X_H^{ref}
$1.7 \times 10^4/p(\text{Torr})$	5×10^5	0.01	1	0.1	0.1
[159]	[159]			[162]	†

† Equilibrium calculations for $T_f = 2400°C$ (cf. Fig. 3.2).

For reactors with $Pe \ll 1$, the diffusion length can be calculated from the diffusion equation; it depends on the geometry of the reactor alone. In the case of HFCVD, l_d can be estimated (with the filament radius r_f and the filament–substrate distance d_f) as $l_d = d_f \ln(d_f/r_f)$; for the hydrogen mole fraction at the surface it follows that

$$X_H^s = \frac{4D_H}{\bar{v}_H d_f \ln(d_f/r_f)\gamma_H} X_H^{ref}. \tag{3.16}$$

With the parameters compiled in Table 3.4, one obtains for a typical HFCVD process with a pressure of 26 mbar, a substrate temperature of 900°C and a filament temperature of 2400°C,

$$X_H^s = 0.015 X_H^{ref} = 1.5 \times 10^{-3}, \tag{3.17}$$

which is equivalent to an absolute concentration of atomic hydrogen of 4×10^{-10} mol/cm^3. This is in good agreement with quantitative measurements of [H•] at the substrate surface [150, 163]; for example, for similar conditions Hsu reported values of $1-2 \times 10^{-3}$ (compare also Fig. 3.13).

Equation (3.16) contains, on the one hand, some parameters, such as p, r_f, d_f and T_f (some of them only indirectly), which in principle can be used to influence the process; on the other hand, most of these parameters can be varied only within restricted limits. The pressure p can hardly be reduced below 10 mbar, whereas the assumed filament temperature of 2400°C represents roughly the upper limit for tungsten filaments. The only parameter which can be changed at least even principle by approximately one order of magnitude is the filament distance d_f. Indeed, some of the author's own investigations show that a reduction of d_f from 5 to 0.5 mm leads to an increase of the growth rate by nearly one order of magnitude [106, 132]. Such a low filament distance results, however, in a strongly inhomogeneous deposition, which could be counteracted only by a dense multifilament arrangement. Owing to the high energy input and the large amount of heat to be removed, such an arrangment presents great technological difficulties.

The situation is similar for the MWCVD technique, which is also diffusion-dominated. The diffusion length l_d is again on the order of some centimeters [159], and the mole fraction of atomic hydrogen at the surface is on the order of about $1–2 \times 10^{-3}$; furthermore, as in the case of HFCVD, no signicant improvement of X_H^s seems to be possible.

For convection-dominated reactors, on the other hand, the diffusion length l_d is given by the thickness of the boundary layer, which is on the order of some millimeters or even lower. However, the situation is complicated by the fact that in contrast to diffusion-dominated reactors, homogeneous gas phase reactions (direct recombination, reactions with methyl radicals) represent an additional path for the loss of atomic hydrogen. Goodwin [159] showed that for convection-dominated reactors the transport of atomic hydrogen is much more prone to variation and, as a consequence, to optimization (compare Fig. 3.13); the expense of the gas phase activation required for large-area deposition, and also of the required gas flows, however, is immense ([159]; compare [85]). A detailed presentation of these aspects is, however, beyond the scope of this book.

Finally, it should be noted that even for HFCVD high gas velocities can be obtained by forced convection. Kröger et al. have shown that – in accordance with the above considerations and also with Fig. 3.13 – these conditions result in increased concentrations of atomic hydrogen at the surface and, consequently, also to increased growth rates [160].

3.2.4 Surface Processes

The discussion in the previous sections has shown that the composition of the gas phase, especially the ratios of the concentrations of the various carbon species to each other and to the concentration of atomic hydrogen, has great influence on the nature (i.e. the quality) of the deposited diamond films. The growth rates of the layers, on the other hand, are in part determined by the transport processes. The decisive processes for the nature, quality and growth rates of the deposited films, however, are the processes taking place on the surface of the growing films. The drastic influence of the substrate temperature on the layer quality as well as on the growth rates (see Sect. 3.2.2 and pages 72f) alone calls for this conclusion. These surface processes are discussed in the following at some length.

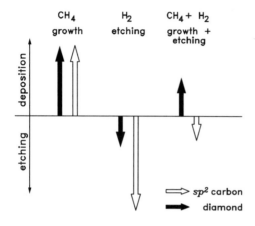

Fig. 3.10. Model of selective etching by atomic hydrogen during diamond deposition

A first attempt to explain diamond growth was proposed by the Russian group of Deryagin [33, 83, 84, 117, 164, 165]. Owing to the very small energy difference between diamond and graphite (Fig. 2.2), there is a certain probability that during a deposition process, besides the stable graphite the metastable diamond is also obtained (Fig. 3.10). However, the etch rate of graphite (or, more generally, of sp^2-bonded carbon) by atomic hydrogen at the temperatures used during diamond deposition is much higher than that of diamond. During deposition in the presence of large concentrations of atomic hydrogen, any sp^2 material obtained is etched back almost completely; thus, in total, the deposition of diamond results (Fig. 3.10)[18].

Indeed, a selectivity of the etch rates of diamond and sp^2-bonded carbon by atomic hydrogen exists, as has been shown by Vietzke et al. [167][19], among others. Therefore it is generally accepted that selective etching of sp^2-bonded carbon plays an important role in the nucleation as well as in the growth of diamond. Also, the model outlined above is capable of explaining a vast

---

[18] In fact, in the early Russian experiments on the deposition of diamond, the back-etching of the obtained sp^2 material by atomic hydrogen was performed not during the deposition itself but during sequential etch steps. Simultaneous back-etching, i.e. the use of large concentrations of atomic hydrogen during deposition, can be regarded as the decisive breakthrough towards low-pressure synthesis of diamond [33, 84, 166].

[19] According to [167], the etch rate of crystalline graphite is higher by one order of magnitude and that of amorphous sp^2 carbon by several orders of magnitude than that of diamond.

variety of observations concerning diamond growth, at least phenomenologically. Nevertheless, it is also generally accepted today that the mechanism of selective etching is not the most important task atomic hydrogen has to perform during diamond growth. The most important processes are, rather, the stabilization of diamond surfaces and the creation of free growth sites on these surfaces by atomic hydrogen.

General Aspects. The surface processes responsible for the growth of diamond are understood in principle, even if the exact details are still unknown to a large extent. Diamond grows at the substrate surface by sequential attachment of well-defined, invariant species, in which process atomic hydrogen fulfills several important tasks. The precise nature of these growth species is still unknown although the majority of authors assume that, at least in the cases of MWCVD and HFCVD, the growth takes place by the attachment of methyl radicals. It cannot, however, be excluded that the decisive species are different for the various crystallographic planes, or that there are differences between those techniques working at atmospheric pressure, at the one hand, and MWCVD and HFCVD, on the other hand.

Despite all these open questions, a generally accepted conception has emerged during recent years (e.g. [54, 55, 88, 162]) which is well-suited to explain diamond growth at least qualitatively. It allows even a quantitative description by means of so-called 'reduced' models, which are discussed in detail below.

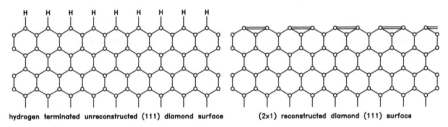

Fig. 3.11. Schematic representation of a unreconstructed, hydrogen-terminated surface and a reconstructed pure diamond {111} surface (modified from [54])

First, however, the nature of diamond surfaces under the special conditions of low-pressure diamond synthesis has to be addressed. Figure 3.11 shows, schematically, on the right-hand side a clean {111} surface. Every surface atom has one free (dangling) bond which is energetically unfavorable; the surface therefore reconstructs, forming π bonds. Thus the surface bears more resemblance to graphite than to diamond and forms, therefore, no convenient starting point for the growth of diamond.

In the presence of atomic hydrogen, however, the {111} surface is completely hydrogen-terminated at the substrate temperatures typical for diamond deposition (Fig. 3.11, left-hand side). The dangling bonds stick out

66   3. Growth Mechanisms

perpendicularly from the surface and are saturated with hydrogen atoms. For the {110} and {100} surfaces the situation is similar. In the case of {100} surfaces, however, which possess two dangling bonds per carbon atom, a complete hydrogen termination (dihydrides) is sterically unfavorable [88]; thus a surface reconstruction by the formation of monohydrides or a mixture of monohydrides and dihydrides is very probable[20].

Thus, in the case of {100} surfaces even the exact nature of the growth surface is still unknown. Although in addition the precise growth species cannot be identified definitively, a modeling of diamond growth is nevertheless possible even quantitatively.

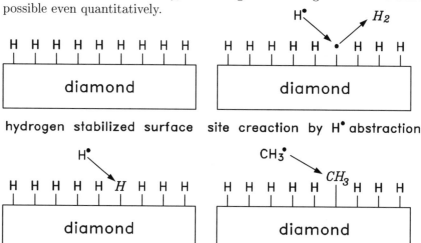

Fig. 3.12. Growth mechanisms of diamond films (schematically, according to [54]). (a) Completely hydrogen-terminated diamond surface. (b) Creation of a growth site by hydrogen abstraction. (c) Annihilation of a growth site by attachment of a hydrogen radical. (d) Diamond growth by attachment of hydrocarbon species

Figure 3.12 shows the essential steps of diamond growth according to the generally accepted picture. The starting point is a completely hydrogen-terminated surface. By reaction with atomic hydrogen (hydrogen abstraction), which is present abundantly in the gas phase, a free site is created on the surface. In almost all cases this growth site is filled by a hydrogen atom again. There is, however, a certain probability for the attachment of a hydrocarbon species (this requires either a radical such as $CH_3^\bullet$ or an unsaturated compound such as C_2H_2; in Fig. 3.12, as an example, the attachment of a methyl radical is shown)[21]. This step therefore represents the actual growth

---

[20] For the nature of diamond surfaces and their various reconstructions in vacuum as well as in the presence of atomic hydrogen, see e.g. [122, 168, 169].

[21] If one assumes that the energy input of at least 10 kWh/carat which is required for the growth of diamond (Table 3.1) is totally consumed by the dissociation of

process in the frame of this model, which in the following is discussed in some more detail.

'Reduced' Growth Models. A complete discussion of diamond growth under low-pressure conditions requires one to take into account several hundreds of possible (surface) reactions. The rate constants of these reactions are unknown to a large extent, which renders a quantitative modeling impossible, quite apart from the complexity of the task. It is possible, however, to reduce the vast number of reactions to a limited number of basic types. With these basic reactions, 'reduced' models can be developed which allow predictions concerning, for example, the parameter dependences of the growth rate or film quality and which can thus be compared with experimental data. Several such reduced models have been published in the literature (e.g. [88, 143, 162]), which are distinguished particularly by the nature of the step of carbon incorporation. As an example, the model proposed by Goodwin [162] will be discussed in some detail in the following. It consists of four basic steps.

1. Formation of an Equilibrium Concentration of Free Growth Sites. The concentration of free surface sites is determined by two reactions, in each of which atomic hydrogen plays the decisive role: the abstraction of a hydrogen atom from a surface carbon atom C_dH (Fig. 3.12b) and the refilling of a free surface site $C_d^\bullet$ (Fig. 3.12c):

$$C_dH + H^\bullet \xleftrightarrow{k_s} C_d^\bullet + H_2 \;, \tag{3.18}$$

$$C_d^\bullet + H^\bullet \xrightarrow{k_f} C_dH \;. \tag{3.19}$$

Equation (3.18) is reversible, whereas the reversal of (3.19) (thermal desorption of hydrogen), on the other hand, is extremely improbable. A closer analysis [162] shows that, if the concentration of atomic hydrogen in the gas phase at the surface is sufficiently high ($X_H = [H]/[H_2] \geq 1 \times 10^{-4}$)[22], the following expression for the equilibrium concentration of free surface sites holds:

$$f^\bullet = \frac{k_s}{k_s + k_f} \;. \tag{3.20}$$

This means that $f^\bullet$ depends only on the substrate temperature but not on the composition of the gas phase. It is a great problem, however, to find data

---

molecular hydrogen, and further, that the atomic hydrogen in turn is consumed by the creation of free surface sites only, a probability for the attachment of a hydrocarbon species at a growth site of only 10^{-4} results [83, 152], in agreement with typical rate constants. This step, therefore, is one of the reasons for the inefficiency of diamond growth.

[22] This requirement is fulfilled for typical HFCVD/MWCVD conditions (and thus all the more for methods with much higher gas phase temperatures); in the following it is therefore assumed that the concentration of free surface sites has assumed this saturation limit.

for the required rate constants (see also below). The values used by Goodwin yield, for 1200 K, a relative concentration of free surface sites of $f^\bullet = 0.14$.[23]

In addition, the combination of reactions (3.18) and (3.19) provides a mechanism for the recombination of atomic hydrogen in which the surface acts as a catalyst (sum of the two reactions: $2\text{H} \longrightarrow \text{H}_2$). This atomic hydrogen has to be redelivered from the gas phase by either diffusion or convection in order to provide a constant $\text{H}^\bullet$ concentration at the surface.

2. Attachment of Reactive Carbon Species to Free Surface Sites. As already mentioned above, either radicals (e.g. $\text{CH}_3^\bullet$) or unsaturated molecules (e.g. acetylene) have to be considered as the growth species:

$$\text{C}_\text{d}^\bullet + \text{C}_n\text{H}_m \xrightarrow{k_\text{a}} \text{C}_\text{d}\text{A} \; . \tag{3.21}$$

In this formulation, the nature of the attached species is still kept open. When selecting the rate constant k_a, however, Goodwin explicitly assumes $\text{CH}_3^\bullet$ as the growth species.

3. Back-Transfer of Adsorbates into the Gas Phase. This step can take place either by thermal desorption or by reaction with atomic hydrogen (etching):

$$\text{C}_\text{d}\text{A} \xrightarrow{k_\text{d}} \text{C}_\text{d}^\bullet + \text{C}_n\text{H}_m \; , \tag{3.22}$$

$$\text{C}_\text{d}\text{A} + q\text{H} \xrightarrow{k_\text{e}} \text{C}_\text{d}^\bullet + \text{products} \; . \tag{3.23}$$

These reactions, which are not taken into account in all models, are necessary to explain the growth of well-faceted crystals. This requires the attachment process to take place predominantly at steps which subsequently migrate over the crystal surface. Since surface migration is expected to be negligible owing to the strong C–C bond and the hydrogen termination, species adsorbed on the terraces themselves must not be incorporated but, rather, have to leave the surface again. In addition, the etching step (3.23) is assumed to be of first order in [H].

4. Incorporation of Adsorbates into the Diamond Lattice. Again, this step takes place by abstraction of hydrogen atoms from the adsorbates by atomic hydrogen:

$$\text{C}_\text{d}\text{A} + (m-1)\text{H} \xrightarrow{k_\text{i}} \text{C}_\text{d}\text{H} + (m-1)\text{H}_2 \; . \tag{3.24}$$

For simplicity it is again assumed that this reaction is also of first order in [H]. Nevertheless it is very likely that this step is more complex and consists of several subsequent reactions. Since the incorporation takes place predominantly at steps (see above), k_i should in addition be proportional to the density of these steps, which, however, for the sake of simplicity is regarded as constant.

---

[23] Estimates of $f^\bullet = 0.12$ and $f^\bullet = 0.37$ for 1000 and 1750 K, respectively, are made by Angus et al. [55], whereas Butler and Woodin [88], on the other hand, assume a value of only 0.03 at 1200 K.

Growth Rates. On the basis of reaction mechanism described above, and taking into account the assumptions discussed above for each of the steps, Goodwin derives the following expression for the growth rate of diamond films [162]:

$$R = \frac{g_1 f^{\bullet} [C_n H_m][H]}{g_2 + [H]} \tag{3.25}$$

with $\quad g_1 = \dfrac{k_i k_a n_s}{(k_e + k_i) n_d} \quad$ and $\quad g_2 = \dfrac{k_d}{k_e + k_i}$, $\qquad$ (3.26)

where n_d and n_s are the density and the surface density, respectively, of diamond.

From (3.25) it is evident that two extreme cases are of especial interest: for low concentrations of atomic hydrogen ($[H] \ll g_2$) the growth rate is proportional to $[H]$, whereas it is, on the other hand, independent of $[H]$ for very high concentrations ($[H] \gg g_2$). Both cases can indeed be observed experimentally (compare also Fig. 3.13).

At this point, the problem of the selection of rate constants, already mentioned, has to be addressed again. These are well known for the hydrocarbon gas phase chemistry; for the surface reactions considered here, however, almost no data are available. Therefore in most cases data for gas phase reactions are used (e.g. hydrogen abstraction from alkane molecules) but are, nevertheless, often modified somewhat randomly, e.g. with reference to steric effects.

According to Goodwin it is possible to fit (3.25), which contains three parameters ($f^{\bullet}$, g_1 and g_2), each of which is in turn a combination of several reaction constants, with a variety of sets of parameters to existing growth rate data and also to other, more detailed growth models [162]. Exact fits to specific sets of data are not decisive at this point; rather, it is of great importance that (3.25) reflects, on the one hand, the parameter dependences for a given deposition technique correctly, and that it predicts, on the other hand, the different rates of different techniques in the correct ratio (see Fig. 3.13).

Formation of Defects. The reduced mechanism discussed so far describes only the growth of diamond. The formation of defects, i.e. the incorporation of non-diamond-bonded carbon, has yet not taken into account. According to Goodwin, defects in the form of sp^2-bonded carbon are incorporated into the diamond lattice in those cases where two neighboring surface adsorbates $C_d A$ react with each other instead of contributing to diamond growth by hydrogen abstraction according to reaction (3.24). For the defect generation rate, it follows therefore that

$$\dot{R}_{\text{def}} = k_{\text{def}} [C_d A]^2 , \tag{3.27}$$

which leads to the following expression for the relative defect concentration in the film:

$$X_{\text{def}} = \frac{\dot{R}_{\text{def}}}{\dot{R}_{\text{C}}} \propto \frac{R}{[\text{H}]^2} \tag{3.28}$$

(with $\dot{R} = n_{\text{d}} R$), where the proportionality constant again depends on temperature. This means that for a constant concentration of atomic hydrogen, the quality is inversely proportional to the growth rate[24]. On the other hand, the quality can be improved by increasing the concentration of atomic hydrogen. Both predictions are in good agreement with experimental observations.

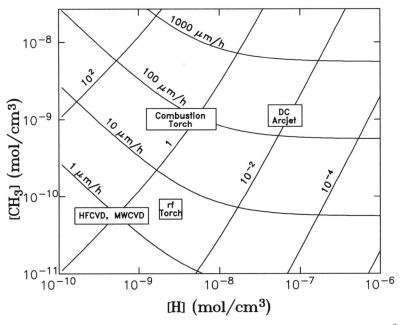

Fig. 3.13. Dependence of the growth rate and relative concentration of sp^2 carbon on the concentrations of hydrogen and methyl radicals ((3.25) and (3.28)) [162]. The sp^2 concentration for HFCVD conditions has been normalized to unity. The working regions of some important deposition techniques are indicated

In Fig. 3.13 the growth rates of diamond films according to (3.25) and their quality according to (3.28) are presented as a function of the concentrations of atomic hydrogen and methyl radicals.

Besides the mechanism due to Goodwin described above, a number of other reduced models of diamond growth exist in the literature, as already mentioned. All have in common the steps leading to the creation of an equilibrium concentration of free surface sites ((3.18) and (3.19)), and the attachment step (3.21). Differences exist, however, concerning the incorporation of surface adsorbates into the diamond lattice (3.24), the importance of

---

[24] This is a phenomenon which is widespread in thin-film technology.

the back-etching step (3.23) and the mechanisms of defect formation (3.27). Without claiming completeness, some of the existing variants are briefly listed in the following:

- According to (3.24), Goodwin assumes that the incorporation takes place via hydrogen abstraction from the surface adsorbates. Butler and Woodin [88] postulate, in contrast, that a carbon atom is incorporated into the lattice only if at least a second bond has been formed between the adsorbate and lattice. Thus the incorporation step requires the reaction of C_dA with neighboring carbon atoms C_dH. An analysis of this mechanism yields the result that the rate of carbon incorporation into the diamond lattice is directly proportional to the concentration of atomic hydrogen [H] (the dependence on the concentration of precursors is not discussed explicitly).
- Evans and Angus [143] assume that for high concentrations of growth species a saturation of the surface with adsorbates occurs, which finally renders the growth rate independent of the concentration of precursors. For the growth rate the expression

$$R = \frac{g_3[CH_x]}{1 + g_4[CH_x]/[H]} \qquad (3.29)$$

follows, where g_3 and g_4 are again combinations of rate constants which depend only on temperature. The functional dependence of the rate on the concentration of atomic hydrogen is the same as that given by Goodwin (3.25); in particular, in both cases a saturation of the rate follows for large [H] – in contrast to the prediction of Butler and Woodin. A difference can be found in the dependence of the rate on the concentration of precursors: according to Goodwin, it is always linear, whereas Evans and Angus predict in this case also a saturation effect, which is indeed observed experimentally (e.g. [170, 143, 171])[25].
- Angus et al. [158] assume that sp^2 defects C_{sp^2} are formed in those cases where a surface adsorbate C_dA reacts with hydrocarbon species; nevertheless, such defects can be etched back by atomic hydrogen:

$$C_dA + C_xH_y \longrightarrow C_{sp^2}, \qquad (3.30)$$
$$C_{sp^2} + H^\bullet \longrightarrow CH_4, C_2H_2. \qquad (3.31)$$

An analysis of this reaction scheme again yields (3.28) for the concentration of sp^2-bonded carbon, which has also been derived by Goodwin[26].
- Butler and Woodin [88] distinguish between sp^3 defects, which are caused by incomplete hydrogen abstraction and which are overgrown by subsequent layers, and sp^2 defects, which are formed by attachment of a hydrocarbon species to a growth complex $C_dA\cdot\cdot C_dH$. Without going into

---

[25] However, Goodwin explicitly excluded from his considerations the conditions leading to this saturation.

[26] In [142] the more general expression $X_{def} \propto R/[H]^n$ can be found.

details, it can be stated that the results of an analysis of the model of Butler and Woodin yields that the quality of diamond layers increases with [H] but decreases with [C_xH_y].

These variants have been presented here in order to demonstrate that even for these reduced growth models, distinct differences exist in the literature; this is mainly caused by the fact that the detailed microscopic growth steps which are addressed below are still unknown to a large extent. On the other hand, some of the reduced models listed above are just special cases of more general considerations of other authors; in addition, almost all models yield rather comparable results. This is mainly due to the facts that all models have steps (3.18)–(3.21) in common, that the actual incorporation step involves in each case atomic hydrogen and that it is assumed in each case that the respective reaction is of first order in [H].

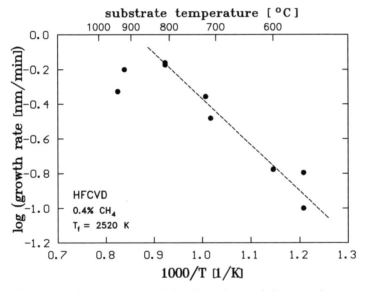

Fig. 3.14. Arrhenius plot of the dependence of the growth rate on the substrate temperature (according to [106])

Influence of Substrate Temperature. Finally, the influence of the substrate temperature T_s on diamond growth, which is inherent in the above equations in the (combinations of) rate constants, has to be discussed. Figure 3.14 shows a set of data from the author's own HFCVD experiments which is typical of our own results as well as for the majority of the investigations published in the literature (e.g. [172, 173]). The growth rate increases with increasing T_s; in Arrhenius plots, straight lines result in most cases. At temperatures between 900 and 1100°C, however, a decrease of the rate is observed which is accompanied by a drastic degradation of the diamond quality; Raman spectra hint at an increasing fraction of graphite within the films. At low temperatures also, an increasing degradation of the quality is observed, which, however, is not caused primarily by the deposition of graphite but, rather, by an increasing fraction of amorphous carbon. Finally, it has to be mentioned that the activation energy resulting from the Arrhenius analysis[27] and the transition temperature to the graphite region depend on the deposition technique, and for a given method also on the process parameters.

These experimental observations concerning the influence of the substrate temperature can be explained rather easily on the basis of the mechanisms discussed so far. According to (3.25), the growth rate depends on T_s via the three parameter combinations $f^\bullet$, g_1 and g_2. Especially, the number of free surface sites increases with temperature [88], which explains the increase of the rate with T_s. However, since the importance of the three parameters varies with the gas phase composition it is comprehensible that the observed 'effective' activation energy depends on the method and the parameters used.

The graphitization at high temperatures follows from the simple model of Sommer and Smith presented in Sect. 3.2.2 (see Fig. 3.3); this model also predicts the dependence of the transition temperature on the gas phase composition. The decrease of the rate at high temperature follows inevitably from the advancing graphitization since sp^2 carbon is etched much more strongly by atomic hydrogen than diamond. The degradation of the diamond quality at low temperatures can be explained in the frame of Goodwin's model by the decrease of the desorption and back-etch processes (3.22) and (3.23). This decrease of the quality at low T_s is not predicted by the model of Sommer and Smith; it has to be taken into account, however, that this degradation is primarily caused by the incorporation of amorphous carbon, a modification not considered in the model of Sommer and Smith.

Detailed Mechanisms. Although the reduced models of diamond growth presented above are capable of describing the growth process qualitatively (and to some extent even quantitatively), it has to be pointed out that the detailed mechanisms, i.e. the attachment of well-defined growth species to well-defined growth sites, are still open to discussion. Up to now it has not been possible either to identify the growth species unequivocally or to characterize all growth surfaces in detail. Quantitative modeling is, in addition, also prevented by the fact that the rate constants for the various possible surface reactions are unknown. For reasons of concentration, only CH_4, $CH_3^\bullet$ and C_2H_2 come into consideration as growth species[28] [55, 83, 164]. However, CH_4 molecules are very stable; the discussion in the literature thus

---

[27] Approximately 0.6 eV for the data presented in Fig. 3.14. The values published in the literature scatter between 0.5 and 1.5 eV.

[28] For all other species the gas phase concentrations are – at least for HFCVD and MWCVD – too low to explain the observed growth rates; for methods working

concentrates on $CH_3^{\bullet}$ and C_2H_2; at the present time $CH_3^{\bullet}$ is, almost generally, assumed to be the actual growth species. Clear hints in this direction can also be found in our experiments presented above, from which it is evident that for HFCVD as well as for MWCVD the growth rates correlate with the concentration of monomeric carbon species, whereas on the other hand the quality correlates inversely with the concentration of carbon species. However, it cannot be excluded that for different growth surfaces the attachment step involves different growth species [164, 174] (see also Sect. 3.2.5). In addition the possibility exists that the growth rates are not determined by the actual attachment process but, rather, by the nucleation of new monolayers [174] or even by the diffusion of growth sites on the surface [176].

A variety of detailed models of diamond growth has, meanwhile, been published in the literature (e.g. [76, 88, 164, 174]). However, none of them has been verified unequivocally, either experimentally or even by simulations, up to now. Therefore – and also for reasons of space – we refrain from a further presentation of these models.

Finally the role of oxygen (species) in surface mechanisms during diamond growth has to be addressed briefly. The discussion in Sect. 3.2.2 revealed, on the one hand, that the primary task of oxygen during diamond deposition is the regulation of the concentration of free carbon in the gas phase, but also, on the other hand, that the surface processes are influenced by the presence of oxygen. Nevertheless, as in the oxygen-free case, the detailed mechanisms are again unknown to a large extent: in the literature it is generally assumed that OH species – similarly to atomic hydrogen – selectively etch sp^2-bonded carbon; in addition, the stabilization of surfaces by OH species is also discussed [139, 142, 147, 164, 177]. However, the mechanisms proposed up to now are rather speculative since quantitative analyses are not available at the present time.

3.2.5 Macroscopic Diamond Growth

Whereas – as we have seen – the microscopic details of diamond growth are still unknown to a large extent, some important macroscopic film properties which are relevant to applications, such as morphology and texture (and even optical quality) can now be adjusted reproducibly. In most cases, polycrystalline diamond films possess a fiber texture, i.e. all crystallites have one common crystal axis normal to the surface, whereas for the other axis no preferred orientation exists. By variation of the process parameters (carbon concentration, substrate temperature and degree of excitation), the texture can be adjusted reproducibly from $\langle 111 \rangle$ through $\langle 110 \rangle$ to $\langle 100 \rangle$ (Figs. 3.15 and 3.16) [178, 179, 180, 181, 182]. The texture is, in general, more pronounced the thicker the layer.

---

at very high gas phase temperatures, such as thermal plasmas or plasma jets, species such as C_2, C_2H and C are also discussed [174, 175].

3.2 Growth of Diamond Films   75

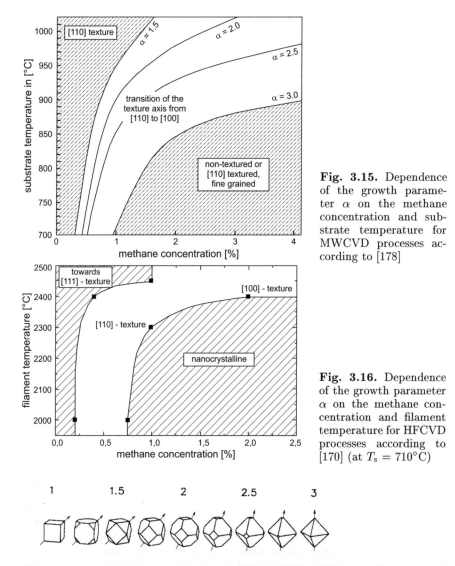

Fig. 3.15. Dependence of the growth parameter α on the methane concentration and substrate temperature for MWCVD processes according to [178]

Fig. 3.16. Dependence of the growth parameter α on the methane concentration and filament temperature for HFCVD processes according to [170] (at $T_s = 710°C$)

Fig. 3.17. Idiomorphic crystal shapes for different values of the growth parameter α. The *arrows* mark the largest diameter, i.e. the direction of fastest growth [178]

The development of the fiber texture can be explained by means of the model of evolutionary crystal growth (van der Drift model) [182, 183]. In isolated crystals the direction of fastest growth is parallel to the largest diameter of the crystal (Fig. 3.17). The shape of the crystals is therefore determined by the growth parameter [182]

$$\alpha = \sqrt{3}\frac{v_{100}}{v_{111}}, \tag{3.32}$$

which describes the ratio of the growth velocities v_{100} and v_{111} on {100} and {111} surfaces, respectively. In most cases, diamond growth starts from isolated, randomly oriented crystallites (Sect. 4.2.1, but see also Sect. 4.3); with increasing film thickness, those crystals for which the direction of fastest growth is perpendicular to the surface will overgrow all other crystals more and more (see Fig. 4.8).

From Figs. 3.15 and 3.16 it is evident that α and thus at least one of the growth velocities depend on the gas phase parameters ([CH$_4$], T_G) as well as on the surface parameters (T_s). α and in turn v_{100} are (at least relatively) the higher

- the higher the methane concentration of the gas phase [178];
- the lower the substrate temperature [178];
- the lower the gas phase temperature [170].

These observations are in good agreement with direct measurements of the growth velocities on the individual surfaces [184, 185]. These measurements show that for all conditions the growth velocity is the highest for {110} surfaces [184]; accordingly, {110} faces were never observed during diamond CVD[29]. The ratio v_{100}/v_{111}, on the other hand, is a complex function of the parameters methane concentration and substrate temperature.

The dependence of the growth parameter on the process parameters has now been established experimentally (Figs. 3.15 and 3.16), thus enabling the deposition of diamond films with the desired texture and also with desired morphology [178, 182]; a microscopic explanation of the parameter dependences of the individual growth velocities, however, has not yet been found. One possibility is that on different surfaces different species are responsible for the growth (compare Sect. 3.2.4); on the other hand, the different growth velocities are also in agreement with the assumption of only one growth species for the various surfaces since, according to Butler and Woodin [88], on {111} surfaces three atoms are necessary for the nucleation of a new monatomic layer, whereas on {110} surface two atoms and on {100} surfaces only one atom are required. Depending on the ratio of the concentrations of atomic hydrogen and the growth species at the surface and (via the migration velocity) also depending on the substrate temperature, different surfaces could be favored for different conditions. However, the observed growth velocities and their parameter dependences do not agree with the trend proposed by Butler and Woodin.

New investigations of Locher et al. [186] show, in addition, that in fact the situation is even more complex. According to this study, the texture and morphology are to a large extent codetermined by the nitrogen concentration in the gas phase[30]. The transition to the ⟨100⟩ texture at high methane

---
[29] The fastest-growing surfaces vanish during the growth of a crystal, see Fig. 3.17.
[30] N_2 is the main impurity in commercially available hydrogen; a concentration of 10 ppm of N_2 in H_2 leads to an N_2/C ratio of about 1000 ppm in the gas phase.

concentrations apparent in Fig. 3.15 can be observed only for nitrogen concentrations > 20 ppm. These results show that the addition of nitrogen apparently influences the growth velocities on {100} and {111} surfaces. Further, it is worth mentioning that nitrogen is preferentially incorporated into {111} faces. A similar influence of nitrogen on the individual growth velocities has also been observed by Jin and Moustakas [187][31]. This effect of only minor traces of contaminant atoms on the growth velocities of the individual diamond surfaces is a clear hint that the attachment mechanisms during diamond growth seem to be far more complex than have been assumed in the various models published up to now (see Sect. 3.2.4).

It can be stated therefore, as a summary, that macroscopic properties of polycrystalline diamond films such as texture and morphology can be controlled technologically, thus enabling their reproducible adjustment. However, the underlying mechanisms are by no means understood today; in order to elucidate these mechanisms a detailed understanding of the attachment processes and, therefore, a knowledge of the growth species for the individual crystal faces are required. On the other hand, the experimentally observed parameter dependences of α can contribute to clarify these mechanisms.

3.2.6 Conclusions

The preceding brief presentation of the present state of knowledge of the mechanisms of low-pressure synthesis of diamond has revealed that simple thermodynamic arguments allow a preliminary description, but this is not sufficient for quantitative considerations. A more detailed discussion requires the separation of gas phase, transport and surface processes. In the gas phase the radical species which are responsible for the growth of diamond are created; the underlying processes are understood quite well. The concentrations of radicals at the surface, which are decisive for the growth rates and the film quality, are determined by transport processes; for these processes an understanding has also been developed to an extent which allow quantitative considerations. However, the processes taking place at the surface are decisive; a detailed description of these processes has turned out to be impossible up to now. There are two reasons for this. On the one hand, the precise natures of diamond surfaces under the conditions of diamond growth are not known for all principal surfaces; on the other hand, the rate constants for surface reactions are unknown to a large extent. Despite these open questions, reduced models can be derived which describe the growth of diamond qualitatively or even semiquantitatively very well, although up to now neither the growth surface (i.e. the growth sites) nor the growth species has

---

[31] The presence of boron in the gas phase also influences the texture and morphology [95, 188]. However, its effect is quite contrary to that of nitrogen: boron is incorporated preferentially into {100} faces; in its presence, ⟨110⟩ and ⟨111⟩ textures are expecially favored [95].

been identified unequivocally. These reduced models enable the development of strategies to improve the quality of diamond films and to increase their growth rates (see Fig. 3.13).

In addition, the presentation in Sect. 3.2.5 has shown that – although the detailed mechanisms of low-pressure synthesis of diamond are still not understood – some important properties such as texture, morphology, surface roughness and the concentration of defects can be predicted by means of empirical relationships with great precision and be adjusted reproducibly.

It has to be pointed out, however, that any further optimization of diamond layers with respect to special properties or special applications requires a detailed understanding of the specific nature of diamond growth (the thermal conductivity or the avoidance of certain defects may serve as examples). On the other hand, it has to be stressed that the quality of polycrystalline diamond films (PCDs) achievable at the present time is quite sufficient for a variety of applications. If, despite this fact, PCDs failed by far to find the broad technological applications which were predicted by the almost euphoric market analyses at the beginning at the 1990s[32], then this is not a question of the quality of polycrystalline diamond layers achievable at the present time but, rather, a question of their production costs[33]. The costs of diamond deposition, however, are to a large extent determined by the mechanisms described above (provision of a high gas phase temperature over an area as large as possible, low degree of utilization of atomic hydrogen, low degree of utilization of the carbon precursors); it seems therefore that, according to the present state of knowledge, a drastic reduction of the production costs cannot be reached easily.

It is, finally, evident from the presentation in the preceding sections that the *growth* of diamond – the nucleation will be discussed separately in Sect. 4.2 – relies almost exclusively on chemical mechanisms. In fact, in some methods such as plasma-assisted methods, physical processes such as electron impact dissociation of hydrogen also play a role; nevertheless their only task is to provide a supersaturation of atomic hydrogen. The essential mechanisms of diamond deposition are of a chemical nature; they rely on gas phase and surface reactions in which atomic hydrogen – serving several tasks simultaneously – plays the decisive role. In sharp contrast, it will become evident in the following sections that the growth of cubic boron nitride – despite all the similarities between diamond and c-BN discussed in Sect. 2.2.2 – depends almost exclusively on physical mechanisms.

---

[32] Compare, for example, [189] and the analyses cited therein.

[33] For reasons of space we refrain from a more comprehensive discussion of this aspect; for more details, the reader is referred to [190] for example.

3.3 Growth of c-BN Films

3.3.1 Methods and Parameters

The growth of c-BN films has been investigated comprehensively by the author's group. The starting point was a data collection concerning the deposition of c-BN based on an intensive literature survey [72, 96, 191, 192]; according to the results of this literature search the methods for which successful c-BN deposition has been claimed can be divided into three groups (Fig. 3.18): ion-assisted PVD methods, ion-assisted CVD methods and CVD methods working without ion bombardment ('chemical' CVD methods).

ion-assisted PVD	ion-assisted CVD	'chemical CVD'
nanocrystalline c–BN		???
◆ RF sputtering ◆ magnetron sputtering ◆ ion plating ◆ ion-beam-assisted deposition ◆ MSIBD	◆ ECR–CVD + bias ◆ ICP–CVD + bias ◆ RF–PECVD + bias	◆ ICP + filament ◆ plasma jet ◆ DC plasma ◆ microwave CVD
• Rely on the same mechanisms • Fairly well investigated • Fairly well established • First models exist		• Little investigated • Some confusion exists • Mechanisms unknown • Analogy to diamond?

Fig. 3.18. Techniques for the deposition of c-BN. A closer analysis of the literature data, however, reveals that the 'chemical' CVD methods investigated up to now are not suited for c-BN deposition

A closer analysis of the existing data, however, reveals that the 'chemical' CVD methods investigated up to now are not suited for c-BN deposition. The misinterpretations of the relevant authors can be traced back to a poor characterization of the BN films obtained, which in almost all cases relied on infrared spectroscopy alone; it has turned out, however, that contamination as well as stoichiometry problems can lead to absorption in the infrared spectra at the same energy position as that of the c-BN reststrahlen band[34]. This finding caused us to introduce the 'working definition of c-BN films' which was discussed above in Sect. 2.4.3.

In this context it is nevertheless worth mentioning that almost all 'chemical' CVD methods investigated up to now have tried to copy the conditions of diamond deposition, i.e. they worked with high gas phase temperatures and a high concentration of (atomic) hydrogen in the gas phase. The existence of a 'chemical route' to the deposition of c-BN can, nonetheless, not be excluded at the present time; it is, however, evident that processes analogous to

---
[34] B_4C, for example, has an absorption band at 1080 cm^{-1} [99], while B_3N absorbs at 1055 cm^{-1} [193].

diamond deposition are not suited to reach this aim. The underlying reasons will be discussed in Sect. 3.3.6.

As a consequence, according to the present state of knowledge, the only methods suited for the deposition of c-BN are those which work with a strong bombardment of the growing films with ions. The BN modification obtained by such a deposition process depends on the following *internal* process parameters:

- ion energy E_i
- ion angle of incidence Θ_i
- ion flux F_i
- boron flux F_B
- ion mass m_i
- substrate temperature T_s

However, the influences of ion flux and boron flux are not independent of each other; rather, the combined parameter $F = F_i/F_B$ is decisive. This means that the deposition of c-BN is determined by the ion bombardment $\mathcal{B}$ and the substrate temperature T_s (Fig. 3.19). Here, $\mathcal{B} = f(E_i, F, m_i, \Theta_i)$ itself is a complex function of ion energy, ion mass and ion angle of incidence as well as the ion/neutral flux ratio F.

The ion bombardment and the substrate temperature each cause or at least influence a wide variety of processes at the surface of the growing film or in its interior. Those processes which have been claimed in the literature to be of relevance for c-BN deposition are presented schematically in Fig. 3.19. They will be discussed further in Sect. 3.3.3; prior to this it is necessary to quantify the conditions required for the deposition of c-BN in terms of the internal parameters listed above.

However, the internal process parameters that are decisive for c-BN deposition can only be quantified for ion beam techniques such as IBAD (ion beam assisted deposition) or MSIBD (mass-selected ion beam deposition). As a consequence, the next step of our investigation of the growth of c-BN consists of an analysis of all IBAD data existing in the literature[35]. In Figs. 3.20 and 3.21 the BN modifications obtained are shown as a function of the ion/atom flux ratio F, on the one hand, and of the ion energy and the substrate temperature, respectively, on the other hand. In these figures, to the best of our knowledge, all investigations in the literature are taken into account which could be quantified in terms of the internal parameters and in which the deposition of c-BN was proven beyond doubt. All other relevant parameters (ion mass(es), ion angle of incidence, T_s in Fig. 3.20 and E_i in Fig. 3.21) are comparable the experiments included here[36]. In particular, in all investigations an ion beam with a composition of Ar/$N_2 \approx$ 1:1 was used.

Each of the figures shows three well-defined regions: for low ion bombardment (energy as well as flux ratio) only h-BN is obtained (Fig. 3.20). If the

---

[35] MSIBD represents a special case which will be discussed in detail in Sect. 3.3.5.
[36] These parameters, together with further details of the data evaluation, are summarized in [72, 96].

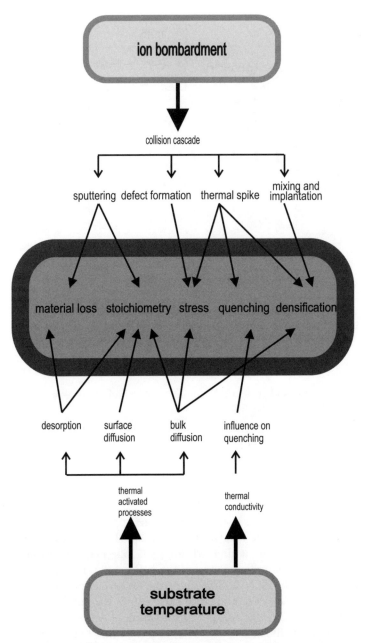

Fig. 3.19. Influence of ion bombardment and substrate temperature on the elementary processes during the deposition of c-BN [97]

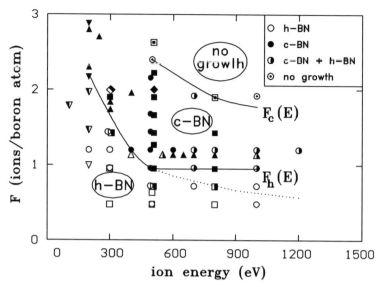

Fig. 3.20. Dependence of the BN modification on the ion energy E_i and the ratio F for IBAD processes. The individual modifications are indicated by the filling of the various symbols; different symbols refer to different authors. For details of the data evaluation, as well as for the corresponding references, the reader is referred to [72, 96]

ion bombardment is increased, the c-BN region is reached. For very strong bombardments, no film growth at all can be observed.

Finally, the range of temperatures in which the deposition of c-BN is possible has to be addressed briefly. Here, a lower limit of $T_{s,\min} \approx 150°C$ seems to exist, in principle, below which c-BN growth no longer takes place [91, 97][37], whereas, on the other hand, no upper temperature limit has been found up to now; the deposition of c-BN is still possible even for $T_s = 1300°C$ [195].

The mechanisms responsible for the various regions apparent in Figs. 3.20 and 3.21 will be discussed in some length in Sects. 3.3.3–3.3.5. At this point, only one secondary effect will be addressed: for high ion energies and high substrate temperatures the boundary between the h-BN and c-BN regions is fixed at a value of F equal to $F_h = 1$. Since in these experiments nitrogen is provided almost exclusively as ions, and since the ion flux is composed of Ar and N_2 with a ratio of 1:1, a value of $F_h = 1$ ion/atom is necessary for the deposition of stoichiometric films. Stoichiometry, on the other hand, is a sine qua non for the growth of c-BN (see below). Thus, for high E_i or a high T_s, F_h is determined by the stoichiometry condition rather than by the influence of the ion bombardment $\mathcal{B}$. This conclusion is confirmed by experiments of

---

[37] Very recent findings of Feldermann et al. seem to indicate that this limit holds for the nucleation of c-BN rather than for its growth [194].

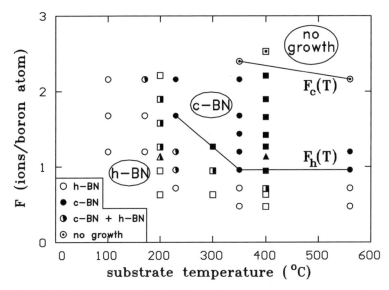

Fig. 3.21. Dependence of the BN modification on the temperature T_s and the ratio F for IBAD processes. The individual modifications are indicated by the filling of the various symbols; different symbols refer to different authors. For details of the data evaluation, as well as for the corresponding references, the reader is referred to [72, 96]

Mirkarimi et al. [196] in which sufficient low-energy nitrogen was provided (dotted line in Fig. 3.20). Under such conditions, values of $F_h < 1$ are indeed observed for high E_i.

At this point of the discussion, the various growth regions apparent in Figs. 3.20 and 3.21 are valid for IBAD processes only, for which all internal process parameters can be quantified. For plasma methods, a quantification of the data in terms of these internal process parameters is not possible, since the flux of boron atoms onto the surface and, as a consequence, the ion/atom flux ratio F cannot be determined even in principle. In order to be able to compare all ion-assisted methods of c-BN deposition, a reparametrization of the results presented in Figs. 3.20 and 3.21 has been carried out; to this end, instead of the ratio of ions to impinging boron atoms F, the ratio of ions to the number of boron atoms incorporated within the films F^* is considered. In the first instance, this step bears two disadvantages. First, the discussion is no longer restricted to internal process parameters; rather, $F^* = F_i/(p_B F_B)$ contains the boron incorporation probability p_B (see below) and thus a process result. In addition, the boundary F_c shifts to infinity since the boron incorporation probability is essentially zero in the 'no growth' region. On the other hand, this operation has the great advantage that F^* can be determined for plasma processes also; as input data, the growth rate, the density of the films, and the plasma density n_e during the process are required. From n_e, using the ion flux j_i onto the substrate can be determined [99]. Likewise, for plasma processes the average ion energy can be calculated from the applied bias voltage V_B and the plasma potential V_P:

$$E_i = k\, e(V_B + V_P), \qquad (3.34)$$

where k is a parameter depending on the pressure which describes the influence of collisions in the sheath.

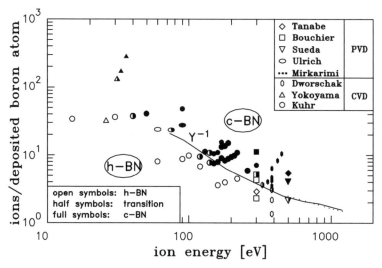

Fig. 3.22. F^*/E_i diagram for the deposition of c-BN. The individual modifications are indicated by the filling of the symbols (note: the 'no growth' region has shifted to infinity); different symbols refer to different authors. For details of the data evaluation, as well as for the corresponding references, the reader is referred to [72, 96]

This results in a parametrization which, in principe, allows one to compare all methods which up to now have been successfully used for the deposition of c-BN. In Fig. 3.22, to the best of our knowledge all experiments for which, on the one hand, all data necessary for quantification have been reported and for which, on the other hand, the deposition of c-BN has been proven beyond doubt are presented in an F^*/E_i diagram. From this figure, which plays a central role in the modeling of c-BN growth (Sect. 3.3.4 and 3.3.5), some important conclusions can be derived:

- In the F^*/E_i diagram, there exists a universal boundary between the h-BN region and the c-BN region which is independent of the method applied.

- c-BN deposition is possible in an energy range from 50 to approximately 1200 eV.[38].
- Low ion energies (> ca. 50 eV) can be compensated by high ion currents (or, better, high F^*) and vice versa.
- There are no differences between PVD and CVD methods in this context.

This means that, for all techniques which can be used at the present time for the deposition of c-BN, the *primary* growth mechanisms have to be the same. In particular, there are no differences between PVD and CVD methods concerning the growth mechanisms. These are dominated by the ion bombardment (and the substrate temperature); chemical effects do not play a role. These primary mechanisms will be discussed in some length in Sects. 3.3.3–3.3.5.

Nevertheless, intensive experimental investigations have shown that besides these *primary* growth mechanisms some *secondary* effects exist which influence the growth of c-BN. This latter term summarizes some mechanisms which, on the one hand, do not actively contribute to the deposition of c-BN (such as the ion bombardment) but which, on the other hand, can prevent the formation of c-BN or at least reduce the quality of the deposited films. Up to now, the following secondary effects have been identified:

- Stoichiometry: the formation of c-BN is only possible for almost stoichiometric layers ($0.9 \leq B/N \leq 1.1$) [96, 198].
- Incorporation of contamination: films with high contents of the cubic phase can be obtained only if their oxygen concentration is low ($\lesssim 3\,\%$) [199, 200]. For too high O contents, the formation of c-BN is suppressed completely. Carbon, on the other hand, can be incorporated in considerably higher concentrations ($\approx 15\,\%$) [199, 200].
- Presence of hydrogen in the gas phase: the presence of hydrogen reduces the maximum c-BN content obtainable; for too high H concentrations, the formation of c-BN is even prevented completely [201, 202, 203, 204][39].

These secondary effects can explain certain differences which have been observed experimentally between CVD and PVD c-BN films [199, 205]. The most important is the maximum c-BN content, which is limited to approximately 80 % for CVD layers [206, 207, 208, 209], whereas with PVD techniques almost pure c-BN films can be deposited [68, 200, 210, 211]. By intensive investigations of the influence of the gas phase concentration of hydrogen [199, 203] and the incorporation of hydrogen within BN films [201, 205], it could be shown that this limitation is mainly caused by the presence of hy-

---

[38] A new investigation by Baranyai shows that at very low ion energies (ca. 40–50 eV) but very high ion currents a direct transition from the h-BN to the 'no growth' region takes place [197]. Very recently, Feldermann et al. were able to extend the energy range up to 2.5 keV [194].

[39] This observation is, of course, of fundamental importance for a comparison of c-BN and diamond growth (see below).

drogen in the gas phase, which is inevitable for CVD processes. For CVD methods, volatile source compounds are required (see the definition of CVD processes at the beginning of this chapter); all precursors which are commonly used for the deposition of c-BN (e.g. B_2H_6, $(BHN(CH_3))_3$) contain some hydrogen. In [203, 205], possible routes to solve this problem and to improve the properties of CVD c-BN films are presented. Nevertheless, we will refrain from a further discussion of this aspect for reasons of space; it is sufficient to point out here that some secondary effects indeed exist having a negative influence on the formation of c-BN but that – and this is by far more important – for all methods, the growth of c-BN relies on the same primary mechanisms.

Finally, some important film properties should be addressed briefly. Since the primary growth mechanisms are the same for all deposition methods, all c-BN films possess – with the exception of the above-mentioned differences between PVD and CVD layers caused by the presence of hydrogen – the same characteristics independent of the technique applied [97]:

- The growth rates of c-BN films are typically of the order of 5 to 10 nm/min. At the present state of technology, they are limited by the obtainable ion currents (Fig. 3.22); for those techniques which are capable of providing extremely high plasma densities ((3.33)), rates up to 50 nm/min are possible even today [212].
- c-BN films are nanocrystalline. The crystallite sizes are typically of the order of 5–20 nm; in some cases, crystallites up to 100 nm in size have been observed.
- c-BN films possess a high compressive biaxial stress, which is typically of the order of 5 to 10 GPa; however, maximum values up to 28 GPa have been observed [213].
- Partly because of this high stress, but also as a consequence of the special structure of the interface between the substrate and c-BN film (see Sect. 4.4), the adhesion of c-BN films on the substrate is in general poor.

These properties, especially the high compressive stress and the poor adhesion, have prevented up to now any industrial application of c-BN films. As a consequence, the author's group has dealt intensively with these problems: in the first step, the specific mechanisms of stress formation and loss of adhesion were investigated theoretically on the basis of our model of c-BN growth which will be discussed below in Sect. 3.3.5; subsequently, on the basis of these considerations, possible routes to solve these problems were developed [72, 214, 215][40]. However, a presentation of these aspects is beyond the scope of this book. Therefore, in the following discussion only some of the mechanisms of the formation of defects which are responsible for the occurrence of stress are taken into consideration.

---

[40] Very recent results confirm the validity of these theoretical considerations [216, 217].

3.3.2 Gas Phase and Transport Processes

Gas phase processes do not play a role during the ion-assisted deposition of c-BN. This can be concluded from the fact that it is of no importance in which form boron or boron-containing species reach the surface. In ion-beam-assisted deposition (e.g. MSIBD [26, 218, 219]), energetic B$^+$ ions are used which have the same energy E_i as the nitrogen ions, i.e. 100–500 eV. In PVD techniques such as IBAD or sputtering, the energy of the boron atoms that reach the surface depends on the method of their creation (laser ablation, ca. 10 eV; sputtering, ca. 3 eV; evaporation, ca. 0.1 eV). In CVD techniques, undefined BN$_x$C$_y$H$_z$ species are involved the exact nature of which depends on the source compounds used; their energy is only slightly above room temperature. For all these methods, however, independent of nature, charge and energy of the boron species, the same c-BN domain exists according to Fig. 3.22 [41].

Transport processes also are of minor importance during c-BN deposition. Indeed, for plasma processes the transport of ions through the sheath leads to the special form of equation (3.33), which couples the ion current to the electron temperature of the plasma; likewise, collisions in the sheath influence the mean ion energy according to (3.34). These processes, nevertheless, are of only minor importance for the mechanisms of c-BN growth. They just influence the coupling between external and internal parameters.

Table 3.5. Process parameters as well as resident times τ and mean free pathes λ for typical diamond and c-BN deposition processes, respectively (ICP, inductively coupled plasma)

		V (litres)	p (mbar)	S (sccm)[†]	τ (s)	λ (mm)[‡]
Diamond	HFCVD	2	26	200	15.4	0.056
c-BN	ICP CVD	2	2×10^{-2}	75	32×10^{-3}	78

[†] 1 sccm = 0.0169 mbar l s^{-1}.
[‡] A temperature of 300 K and a mean radius of the species of 0.8 Å were assumed. The latter is the average value of the radii of C, B and N atoms. Since in diamond deposition molecular species play decisive roles, λ is overestimated by this assumption.

The different importances of gas phase and transport processes for the deposition of diamond and c-BN follow even from the different working pressures of the respective processes. The residence time τ and the mean free path of a species in a deposition reactor are [220, 221]

---
[41] A exception, which will be discussed in more detail below in Sect. 3.3.5, is formed by MSIBD. However, this exception concerns the ion/solid interaction and is not due to gas phase or transport processes.

$$\tau = \frac{pV}{S} \quad \text{and} \quad \lambda = \frac{kT}{\sqrt{2\pi p d^2}}, \tag{3.35}$$

where V is the volume of the reactor, p the working pressure, S the gas flow into the reactor and d the radius of the colliding partners. In Table 3.5, typical process parameters of a diamond HFCVD and a c-BN ICP-CVD set-up, which are both used in our group and which possess similar volumes of the reactor, and the resulting residence times and mean free paths are compared. From these data it is evident that during c-BN deposition, reactions in the gas phase and during the transport to the surface do not play a role.

3.3.3 Ion-Induced Mechanisms

The discussion in the preceding section has shown that the growth of c-BN relies on surface processes only; nevertheless, the term 'surface' includes here not only the outermost layer of atoms but, rather, the range of the penetration depths of the ions, the interaction of which with the surface – as is evident from the compilation of the relevant process parameters in Sect. 3.3.1 and also from Figs. 3.20–3.22 – together with temperature-induced effects cause the formation of c-BN.

An ion impinging onto a solid transfers energy to the electrons and atomic cores of the solid by means of collisions. At high energies inelastic collisions with the electronic system dominate, at low energies elastic core collisions. In the energy range important for the deposition of c-BN ($E_i \lesssim 1.5$ keV), only nuclear energy losses are of interest. As a results of the penetration of ions, atoms of the solid are displaced from their lattice places, form defects or are even able to leave the surface (sputtering). The impinging ions are either backscattered or incorporated within the solid (implantation), or reach the surface again by means of diffusion. Since, especially during the first collisions of a penetrating ion, considerable energies are transferred to the target atoms, the knock-on atoms are also able to cause the processes listed above; as a consequence, a collision cascade with a duration of approximately 10^{-14} to 10^{-13} s is formed (Fig. 3.19).

As soon as the primary ion and the knock-on target atoms are no longer able to displace further atoms, the remaining energy is dissipated predominantly by phonons (thermal spike, ca. 10^{-11} s). In the context of the growth of c-BN and the modeling of this process, two effects of these thermal spikes are mainly of importance. On the one hand, they create a small volume (of radius r) of very high temperature, which is subsequently cooled down to room temperature very rapidly. Seitz and Köhler [222] calculate, for a thermal energy E_{th} which is deposited at time $t = 0$ at the position $r = 0$, the temperature distribution

$$T(r,t) = \frac{E_{th}}{8\pi^{2/3} c\rho (Dt)^{3/2} \exp\left(\frac{r^2}{4Dt}\right)}, \tag{3.36}$$

where c is the specific heat, ρ the density and D the thermal diffusivity. For an energy $E_{th} = 300$ eV, a time $t_{spike} \approx 10^{-11}$ s and a radius $r_{spike} \approx 1$ nm follow, for which the temperature reaches some 1000 K.

One the other hand, during thermal spikes relaxation processes can take place, by which, for example, defects which have been created during the collision cascade can recombine. The number of relaxation processes per spike can be estimated as [222]

$$n_r \approx 0.016 p \left(\frac{E_{th}}{E_0}\right)^{5/3}, \qquad (3.37)$$

where E_0 is the activation energy of the relaxation process and p a constant of the order of unity.

Those interaction processes of an ion and a solid taking place during the collision cascade and the subsequent thermal spike which are important for the deposition of c-BN are presented schematically in Fig. 3.19. The figure distinguishes between primary processes such as sputtering or defect formation, on the one hand, and secondary effects caused by them (e.g. material loss, densification), on the other hand; it should nevertheless be noted that the latter are influenced not only by the ion bombardment but also by the substrate temperature.

3.3.4 Modeling of c-BN Growth

In the literature a variety of models exist for the growth of c-BN (Table 3.6), which in part were developed originally for the deposition of tetrahedral amorphous carbon (see Sect. 3.4.2). They are distinguished by the mechanisms of the formation of sp^3 bonds as well by the role relaxation processes play during c-BN deposition [97, 215]. In addition, they can be separated into two groups, one of them assuming that the formation of c-BN relies on a phase transition sp$^2 \longrightarrow$ sp^3, the other that attachment processes to existing crystallites play the decisive role. All of these models have in common that each of them emphasizes only one aspect of the complex ion/solid interaction sketched in Fig. 3.19 while all others are taken into account at best marginally. Moreover, in most cases these models consider only the formation of the cubic phase; film properties such as crystallinity, stoichiometry, etc. are, with the exception of the high compressive stress, almost always not taken into consideration.

The most important of these models (Table 3.6) will be presented and discussed briefly in the following, so far as they are of interest for the main topic of this book. For a comprehensive presentation of those aspects of the ion/solid interaction which are of importance for the deposition of c-BN, we refer to [72]; a detailed presentation and judgement of the models published up to now in the literature can be found in [97].

The *quenching model* which was at first formulated by Weissmantel et al. [224], assumes thermal spikes to be responsible for the formation of sp^3

Table 3.6. Comparison of the models for c-BN and ta-C growth proposed in the literature, with respect to the mechanisms of phase formation and the influence of relaxation processes [215]. TS, thermal spike; CC, collision cascade; T_s, substrate temperature

Model	Phase formation	Relaxation
c-BN		
Quenching [26]	TS quenching	
Indirect subplantation [223]	CC knock-on implantation	TS $\longrightarrow sp^2$
Dynamic stress [196]	CC interstitial formation	$T_s \longrightarrow sp^2$
Sputtering [96]	attachment	TS $\longrightarrow$ crystalline c-BN
ta-C		
Quenching [224]	TS quenching	
Direct subplantation [225]	CC penetration	TS $\longrightarrow sp^2$
Static stress [226]	CC interstitial formation	TS $\longrightarrow sp^2$

bonds. These lead, according to (3.36), to conditions equal to those of HPHT processes; as a consequence, sp³ bonds are formed within the spikes. Since cooling from these high temperatures takes place extremely rapidly (10^{14}–10^{15} K/s, quenching), the structures formed during the spikes are 'frozen-in' to a considerable extent [26, 73]. Quenching processes usually lead to metastable, amorphous structures. In addition, the dimensions of thermal spikes (ca. 1 nm, see above) are typically one order of magnitude smaller than the experimentally observed diameter of c-BN crystallites (10–100 nm, see above). Hofsäss et al. [26, 73] attribute the crystallinity and the crystallite size of c-BN films in part to the influence of the substrate temperature[42] and in part to the high ionicity of the B–N bond[43]. They point out further that thermal spikes cool down from the outside to the interior; therefore the bond structure of the environment is adopted during cooling [222]. This description thus requires an independent nucleation of c-BN. Finally, it has to be mentioned that a quantitative formulation of the quenching model has not been presented up to now[44].

---

[42] High temperatures reduce the thermal conductivity and thus also the quenching rate. In addition, the mobility of film-forming species is increased.

[43] A high ionicity of the bond increases, on the one hand, the resistance of a crystalline solid against amorphization by ion bombardment and facilitates, on the other hand, the formation of crystals during ion-assisted deposition. This aspect will be addressed in detail in Sect. 3.3.6 and 3.6.

[44] Very recently, Hofsäss et al. published an extensive quantitative description ('cylindrical spike model') [227], which, however, mainly concentrates on the deposition of ta-C by means of ion beam deposition. A critical judgement on

A second group of models assumes the collision cascade at the beginning of the ion/solid interaction to be responsible for the phase formation. The cascade leads to the generation of defects (mainly interstitials and Frenkel pairs). However, the mechanisms by which these defects finally cause the formation of sp³ bonds are different for the various models. Nevertheless, they all have in common the assumption that the thermal spikes following the collision cascades lead, according to (3.37), to defect recombination; they are therefore detrimentral to the formation of sp³ bonds. In each case, the formation of c-BN is described quantitatively by balance equations taking into account the generation of defects by the ion bombardment and the relaxation of defects by thermal spikes.

The *subplantation model*[45] [209, 223, 229] assumes that, by knock-on implantation (subplantation) of surface atoms into the interior of the film, regions of high density are created locally in which the formation of sp³ bonds is energetically favorable. In fact, in the case of c-BN deposition only the so-called indirect subplantation is of importance, since in almost all cases only nitrogen and argon ions are used (see above). Nitrogen and argon, however, cannot cause a densification of the films since surplus nitrogen (like argon) does not remain within the interior of the films but diffuses back to the surface (see below). Therefore, the decisive quantity is the probability q that an impinging ion subplants a surface boron atom into deeper layers. According to TRIM calculations, q can be approximated by the expression

$$q = 0.054 \left[1 - \exp\left(-\frac{E_i - E_p}{E_q} \right) \right], \qquad (3.38)$$

where E_p and E_q are constants. From the balance equation for defect generation and defect recombination, it follows for the density [97] that

$$\rho = \rho_0 \cdot \left(1 + \frac{q \dfrac{b}{a+b}}{\dfrac{2}{F^*} - q\dfrac{b}{a+b} + 0.0016p \left(\dfrac{E_i}{E_0}\right)^{5/3}} \right). \qquad (3.39)$$

Here, a and b are the rate constants for the recombination and the condensation of interstitials, respectively. Densification by attachment of mobile interstitials (condensation) has been additionally taken into account in an elaboration of earlier formulations of the subplantation model [209, 229, 230] by Robertson [223] in order to be able to explain the existence of c-BN nanocrystallites, since a densification by thermal spikes alone should only lead to amorphous sp³ agglomerates of small dimensions (see above).

---

this work, especially from the point of view of its transferability to standard c-BN deposition, has not yet been given.

[45] The term 'subplantation' (*sub*surface im*plantation*) was first introduced by Lifshitz et al. [228] in order to describe low-energy implantation into the subsurface region.

The *stress models*, on the other hand, are based on the assumption that the large number of interstitials causes mechanical stresses (biaxial stresses)[46] in the films which are high enough that the hydrostatic pressure is sufficiently high to reach the region of typical HPHT processes where the formation of c-BN is energetically favored.

In the literature, two formulations of the stress model exist. The *static stress model* [226, 231, 232] considers the stress present at the end of a thermal spike (frozen-in stress). The volume change and thus the stress are proportional to the number of interstitials generated per impinging ion, n_i; on the other hand, relaxation processes lead to the recombination of defects. Taking both effects together, the relation

$$\sigma \propto \mathcal{E} \frac{n_i}{\frac{1}{F^*} + 0.0016 p \left(\frac{E_i}{E_0}\right)^{5/3}} \tag{3.40}$$

for the stress follows, where $\mathcal{E}$ is the Young's modulus of the film.

In contrast to the static stress model, the *dynamic stress model* of Mirkarimi et al. [196] assumes not that the stress resulting at the end of the deposition is responsible for the phase formation, but rather the instantaneous stress present during the deposition itself. TRIM calculations show that the defects created by the ion bombardment possess a Gaussian local distribution with a maximum at about 0.5–1 nm beneath the surface. With increasing deposition time the surface of the film moves away from any given region in the film; after a certain time, therefore, no more defects are created in this region, whereas on the other hand mechanisms leading to the loss of defects, such as thermally induced recombination, diffusion and condensation are still effective. As a consequence, the instantaneous stress in the vicinity of the surface σ_{inst} which is finally decisive for the formation of c-BN is larger than the static stress. Mirkarimi et al. derive

$$\sigma_{\text{inst}} \propto F \sqrt{E_i m_i} \tag{3.41}$$

for the dependence of σ_{inst} on the most important process parameters.

The high compressive stress of c-BN films is an experimental fact (see above). Nevertheless, the question arises – as is also discussed in [214, 233] and even by the authors of one of the existing stress models [196] – whether this stress is the reason for the formation of c-BN, or only an inevitable consequence of the ion bombardment, or, finally, even a consequence of the formation of c-BN itself. According to (3.40) and (3.46), σ is proportional to the Young's modulus, which is approximately four times higher for c-BN than for h-BN. The steplike increase of the stress when the c-BN region is reached is thus caused by the similar steplike increase of $\mathcal{E}$; it is therefore at least in part also a consequence of the formation of c-BN (see below).

In the context of the stress models, another aspect has to be addressed. At the time of the first formulation by McKenzie et al. [231], it was generally

---

[46] The volume of a Frenkel pair is larger than that of an atom on its lattice site.

assumed that under standard conditions c-BN is only metastable (see Sect. 2.2.2) and that, as a consequence, HPHT conditions are necessary for its formation. According to the new findings presented in Sect. 2.2.2, however, c-BN is the stable modification even at standard condition; thus, following the original argument of McKenzie et al., high pressures and in turn high stresses are not required. On the other hand c-BN is, as compared to h-BN, the denser material; thus at high pressures its formation is favored [232]. In this case the transition from h-BN to c-BN with increasing stress should nevertheless be rather gradual, not abrupt as observed experimentally.

A completely different approach is taken in the *sputter model*, which has been developed by the author's group. It is assumed that growth of c-BN takes place by attachment processes while, simultaneously, the growth of h-BN is suppressed by selective sputtering. The relaxation processes caused by the thermal spikes lead to recombination of defects and thus to an improvement of the film quality. Since the sputter model presents the best description of the existing experimental data it is discussed in the next section in some more detail.

3.3.5 The Sputter Model

The boundary F_c in Figs. 3.20 and 3.21 is, beyond doubt, a resputter limit; i.e. all material deposited is immediately removed by sputtering processes. If one takes into accout additionally the temperature dependence of F_c, which is evident from Fig. 3.21 and which cannot be explained by sputter processes, which are to a first approximation independent of temperature, by means of thermal desorption processes described by the boron sticking coefficient $s_B(T_s)$, one obtains for the flux of boron atoms removed by either sputtering or absorption[47]

$$F_B^R = (1 - s_B) F_B + 0.5 Y_c F_i , \qquad (3.42)$$

where Y_c represents the sputter yield of c-BN. At the boundary F_c, all boron atoms impinging on the surface will be removed at once, i.e. $F_B = F_B^R$. For F_c, therefore, the condition

$$F_c(E_i, m_i, \Theta_i, T_s) = \frac{2 s_B(T_s)}{Y_c(E_i, m_i, \Theta_i)} \qquad (3.43)$$

follows. It is evident from this discussion that F_c is a direct resputter limit, i.e. the growth velocity of c-BN is essentially zero: $v_c = 0$.

The similarities of the parameter dependences of F_c and F_h, which are evident from Figs. 3.20 and 3.21, suggest that F_h is also dominated by sputter processes. This requires a *selectivity of the sputter rates of h-BN and c-BN,*

---

[47] In the following, only boron atoms are taken into account since surplus nitrogen diffuses back to the surface and desorbs. This follows from the fact that under c-BN conditions nitrogen-rich films have never been observed [72]. Likewise, argon ions are not incorporated but leave the film by diffusion and desorption.

i.e. $Y_h > Y_c$. In the first formulation of the sputter model [96, 192, 234] it was thus assumed that F_h is the direct resputter limit of h-BN, i.e. at F_h, $v_h = 0$. This assumption results in a relation for F_h which is analogous to (3.43). This formulation of the sputter model already describes the growth of c-BN qualitatively and, in most cases, even quantitatively [96, 192, 234]. It underestimates, however, the 'width' of the c-BN region F_c/F_h [91].

(Selective) material loss processes such as sputtering or desorption alone, however, cannot explain the growth of c-BN. Rather, the sputter model requires a *preferential bond formation*: on attachment of atoms to c-BN crystallites sp^3 bonds are created, whereas on h-BN crystallites sp^2 bonds are formed. This means, in other words, that the growth of c-BN takes place spontaneously by *attachment processes*[48]. The ion bombardment serves only to suppress the unwanted modification[49].

A closer analysis of the situation, however, taking into account the preferential bond formation, shows that for F_h, i.e. the transition from the h-BN to the c-BN region, it is only required that c-BN crystallites grow faster than h-BN crystallites, i.e. $v_c > v_h$. A complete suppression of the growth of h-BN is thus not necessary for c-BN deposition[50]. Taking into account the different densities of h-BN and c-BN, from the condition of equal growth velocities it follows for the boundary F_h that

$$\frac{s_B - 0.5 Y_c F_h}{\rho_c} = \frac{s_B - 0.5 Y_h F_h}{\rho_h}. \tag{3.44}$$

Thus the sputter model applied to the formulation of the *evolutionary* or *competitive growth* of crystallites yields for the boundary F_h

$$F_h(E_i, m_i, \Theta_i, T_s) = \frac{2 s_B(T_s)}{Y_h(E_i, m_i, \Theta_i)} \frac{\rho_c - \rho_h}{\rho_c - \rho_h/S} \tag{3.45}$$

with the sputter selectivity $S = Y_h/Y_c$.

The boundary F_c still remains a direct resputter limit which is described by (3.43). In this formulation of the sputter model, the width F_c/F_h of the c-BN region (≈ 2.5 at 500 eV, see Fig. 3.20) is predicted correctly by (3.43) and (3.45).

The mechanisms of preferential bond formation and competitive growth of crystallites nevertheless require the existence of c-BN crystallites. The origin of such crystallites, i.e. the nucleation of c-BN, is not described by these mechanisms. The formation of the first c-BN nuclei has therefore to be considered separately; it will be discussed in detail in Sect. 4.4.

---

[48] This can indeed be regarded as an analogy to diamond growth.

[49] This is also in analogy to the deposition of diamond, where selective etching of sp^2 carbon by atomic hydrogen plays a distinct role. Nevertheless, etching takes place chemically in the case of diamond but physically (by means of sputtering) in the case of c-BN.

[50] This is in analogy to the van der Drift model discussed above in the context of texture development during the deposition of diamond.

Finally, it has to be mentioned that the thermal spikes discussed above play a role in the sputter model as well. They are of no importance for the actual formation of c-BN; nevertheless they cause relaxation processes and thus recombination of defects which have been created by the ion bombardment [72, 97, 214][51]. A quantitative consideration of defect creation by ion bombardment and defect relaxation by thermal spikes leads to a model [72, 97, 214], which describes the experimentally observed parameter dependences of the compressive stress of c-BN films correctly; for reasons of space, however, we refrain from a detailed presentation. From the balance equation for the generation of interstitials by the ion bombardment and their recombination according to this model, it follows for the stress that

$$\sigma \propto \frac{\mathcal{E}}{1-2\nu} \frac{n_i}{\frac{2}{F^*} + 0.016p \left(\frac{E_i}{E_0}\right)^{5/3}} \propto \frac{\mathcal{E}}{1-2\nu} E_i^{-2/3}. \qquad (3.46)$$

Two aspects of this equation are of importance at this point. On the one hand, the high Young's modulus of c-BN (800 GPa as compared to 200 GPa for h-BN [72]) causes a stepwise increase of the stress as soon c-BN is deposited instead of h-BN[52]. On the other hand, if all other parameters are kept constant, the stress should decrease with increasing ion energy. Both effects have been recently observed experimentally by several authors [213, 235, 236, 237].

In the following, the major points concerning the growth of c-BN derived in the framework of our model are summarized briefly:

- The existence of c-BN nuclei is a prerequisite; however, at this point no statements concerning the nucleation mechanisms have been made.
- The growth of c-BN proceeds via attachment processes which take place preferentially for the different modifications.
- The growth of h-BN crystallites is suppressed by selective sputtering.
- Thermal spikes cause relaxation processes, which in turn lead to recombination of ion–induced defects (and consequently to stress reduction also).

This list shows that selective sputtering indeed plays a major role in the growth of c-BN but is nevertheless not the only decisive process. The term 'sputter model' may therefore no longer be correct; it will nevertheless be maintained in order to allow an easy identification of the various models.

Verification of the Model. From equations (3.43) and (3.45) some conditions can be derived which must be fulfilled if the growth of c-BN indeed relies on the mechanisms discussed above in the context of the sputter model [234].

---

[51] TRIM calculations show that during c-BN deposition several interstitials per incorporated boron atom are created. Evidently, very effective recombination processes must be involved, which are also favored by the high ionicity of the B–N bonds [214].

[52] This aspect has already been addressed in Sect. 3.3.4 in the context of the stress models for c-BN deposition.

96   3. Growth Mechanisms

1. The energy dependences of the boundaries F_h and F_c should correspond to that of an inverse sputter yield:

$$F_{\mathrm{h,c}}(E_\mathrm{i}) \propto Y_{\mathrm{h,c}}^{-1}(E_\mathrm{i}) \ . \tag{3.47}$$

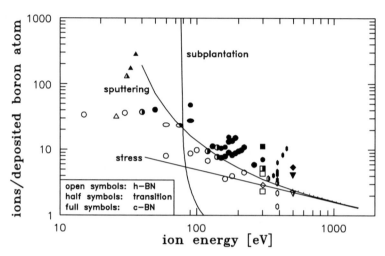

Fig. 3.23. Comparison of the data from Fig. 3.22 with the quantitative predictions of the sputter model, the dynamic stress model (3.41) and the model of indirect subplantation (3.39). For details concerning the choice of parameters see [97, 215]

Figure 3.22 shows that the boundary F_h between the h-BN and c-BN regions can indeed be decribed by an inverse sputter yield[53]. From Fig. 3.23 it is evident that with (3.47), the sputter model gives the best quantitative description of the experimental data for the boundary F_h, out of all the models proposed in the literature[54].

2. The dependence of the boundaries F_h and F_c on the ion angle of incidence Θ_i should, likewise, correspond to that of an inverse sputter yield:

$$F_{\mathrm{h,c}}(\Theta_\mathrm{i}) \propto Y_{\mathrm{h,c}}^{-1}(\Theta_\mathrm{i}) \ , \propto \cos^\gamma(\Theta_\mathrm{i}) \ , \tag{3.48}$$

where $Y(\Theta_\mathrm{i}) = Y(0)\cos^{-\gamma}(\Theta_\mathrm{i})$ has been assumed [91, 238]. The existing data in the literature on the dependence of F_h and F_c on Θ_i can indeed be described satisfactorily by (3.48), as shown in [72, 91]. Large angles of incidence – this

---

[53] Owing to the lack of data concerning the sputter yields of BN films, in Fig. 3.22 data for the system Ar ⟶ Si are shown which have been fitted to the experimental data at 500 eV.

[54] In Fig. 3.23, only the indirect-subplantation and the dynamic stress model have been taken into account. For the quenching model, at present no quantitative formulation exists; for the static stress model, the agreement is even more worse than for the indirect-subplantation model (both models lead to similar parameter dependences, compare (3.39) with (3.40)).

is of importance from the point of view of the modeling of c-BN growth – are thus of advantage for the deposition of c-BN.

3. The dependence of F_h and F_c on the substrate temperature should correspond to that of a sticking coefficient:

$$F_{h,c}(T_s) \propto s_B(T_s) \,. \tag{3.49}$$

All data in the literature (e.g. [206, 239, 240]) show that if all other process parameters are kept constant, the growth rates of BN films decrease with increasing temperature, thus demonstrating the influence of desorption processes. The temperature dependence of the boundaries F_h and F_c is confirmed by Fig. 3.21. In principle, it is possible to switch from the h-BN region to the c-BN region merely by increasing the temperature (compare Figs. 3.21 and 3.29).

4. Owing to the dominance of material loss processes (sputtering, desorption), the boron incorporation probability

$$p_B = s_B - 0.5\, Y_{h,c} F = \frac{F}{F^*} \tag{3.50}$$

should be low. Indeed, within the c-BN regions of Figs. 3.20 and 3.22, $p_B \leq 0.4$ has always been found [72, 234].

5. The absolute sputter yields should be relatively large. From the position of the c-BN region in Fig. 3.20, $Y_c \approx 0.5$–1 can be estimated for 500 eV, for example. By direct measurements of the sputter yields of h-BN and c-BN films and of c-BN crystals, this prediction has been confirmed [72, 234].

6. The sputter yields of h-BN and c-BN must show a selectivity. From the width of the c-BN region in Fig. 3.20, it can be estimated from (3.43) and (3.45) that, at 500 eV,

$$S = \frac{Y_h}{Y_c} = \frac{F_c}{F_h}\left(1 - \frac{\rho_h}{\rho_c}\right) - \frac{\rho_h}{\rho_c} \approx 1.5 \,. \tag{3.51}$$

This prediction was also tested by sputter experiments [72, 234], the results of which are presented in Fig. 3.24 together with the scarce data found in the literature. The figure shows that a selectivity indeed exists; as required from (3.51), it is on the order of[55] $S \approx 1.5$.

The selectivity of the sputter yields of h-BN and c-BN is also confirmed by some results recently published by Westermeyr et al. [242]; these show that the c-BN content in the near-surface region of a mixed c-BN/h-BN layer can be increased distinctly by 1 keV Ar^+ ion bombardment. Since, in contrast, the structure of pure h-BN films is not changed by such an ion bombardment, an ion-induced phase transition cannot be responsible for this effect; rather, this observation can only be explained by selective sputtering.

---

[55] In good agreement with these results are some new data from Baranyai [197], who found for nitrogen and argon ions a sputter selectivity of about 1.4.

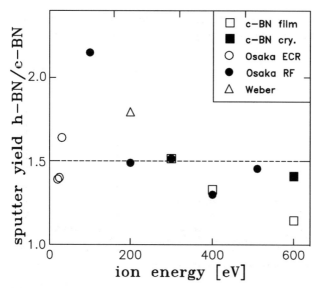

Fig. 3.24. Selectivity of the sputter yields of h-BN and c-BN: compilation of the results of the author's own measurements and the existing literature data [208, 241]

New Results. The data collection concerning the deposition of c-BN which has been discussed in the preceding parts of this chapter and which results in a well defined c-BN region, and the sputter model describing this region form a consistent picture which can be extended – as will be shown in Sect. 4.4 – to the nucleation of c-BN also. In this picture, the data collection and model can be related to a 'standard process' which, despite all differences with respect to the deposition technique, the source compounds and the process parameters (of course only within the limits of the c-BN region), has two important points in common:

- The energy input takes places indirectly, i.e. boron is provided only as low-energy species, whereas the ion flux in most cases consists of equal parts of nitrogen and a heavy inert-gas ion (almost exclusively Ar).
- Nucleation and growth are performed with the same parameters.

For standard processes corresponding to this definition, there are to the best of our knowledge no data deviating from the above description in the literature. Nevertheless, very recently the results of two experiments (Fig. 3.25) have been published which cannot be explained by the sputter model in the formulation presented above. These experiments do not correspond to the standard process outlined above.

- Hofsäss et al. grew c-BN films by means of mass-separated ion beam deposition (MSIBD), where boron as well as nitrogen are provided exclusively in ionic form. This means that the energy input is direct. The

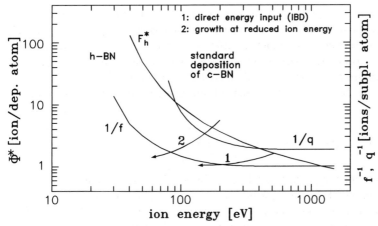

Fig. 3.25. Localization of some new results within the Φ^*/E_i diagram[56]. 1: MS-IBD experiments of Hofsäss et al. [236] with direct energy input. 2: Growth of c-BN outside the c-BN region after nucleation within the region [243]

first experiments seemed to indicate that formation of c-BN is only possible within the c-BN region of the parameter space defined above for experiments with indirect energy input [218]. Recently published results, however, show that in this special case c-BN deposition is also possible below F_h^* (arrow 1 in Fig. 3.25) [236]. A closer analysis of this process reveals that the formation of c-BN – in contrast to the indirect energy input – takes place far from the resputter limit; furthermore, the boron incorporation probability is higher than 70 %. (Selective) sputtering thus seems to play no decisive role in these experiments.

- Hahn et al. [243] describe experiments in which, after nucleation of c-BN above F_h^*, i.e. within the c-BN region, the ion energy (and, as a consequence of the reduced sputtering F^* also) have been reduced (arrow 2 in Fig. 3.25). This experiment led to the growth of almost pure c-BN, whereas without the previous nucleation step, under these conditions only h-BN is obtained. Nevertheless, a threshold energy of about 40–60 eV seems to exist also for this process, below which, even after a preceding nucleation step, no c-BN growth is possible. A similar experiment has also been described earlier by McKenzie et al. [244]; nevertheless, the published details were not sufficient for a definite judgement on these results.

A continuing growth of c-BN crystallites after switching off or reducing the ion bombardment is, in the first instance, not in contradiction with the sputter model. Indeed, it is even required by the concept of competitive growth. Nevertheless, the c-BN crystallites should gradually

---

[56] In order to facilitate comparison with the deposition of ta-C, in the following the ions/total number of incorporated atoms ratio is used: $\Phi^* = F^*/2$.

become overgrown by h-BN, i.e. the c-BN content should decrease with increasing deposition time. This, however, seems not to be the case during the experiments of Hahn et al. and McKenzie et al., although clearly further work is required to establish a complete description of this process.

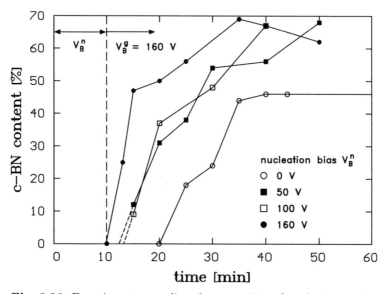

Fig. 3.26. Experiments regarding the separation of nucleation and growth: variation of the bias voltage during the nucleation stage [203, 245]. The figure represents the c-BN content of the actual growth layer without the nucleation layer deposited in the first 10 min of the process [245]

- On the other hand, some of our own experiments performed recently with the ICP-CVD method show that nucleation of c-BN is only possible within the c-BN region of the parameter space [203, 245] (Fig. 3.26). In these experiments, the bias voltage (and thus the ion energy) at the beginning of the deposition, i.e. during the nucleation phase, was varied whereas the following growth step was performed with a bias voltage of 160 V, which was the optimum value for the conditions used. Figure 3.26 shows that only if the nucleation step is performed at 160 V, i.e. within the c-BN region, does c-BN start to grow at the end of the nucleation phase of 10 min. For lower bias voltages during the nucleation step, however, another (bias-dependent) incubation time is observed. These results seem to indicate that the nucleation of c-BN – in contrast to the growth – is possible only within the c-BN region.

For a final classification of these new results, the data existing at present are by no means sufficient. It must nevertheless be pointed out that neither the experiments of Hahn et al. [243] and McKenzie et al. [244] nor that of Hofsäss et al. [236] can be explained in the framework of the present formulation of the sputter model.

The results of Hofsäss et al. show that in c-BN deposition, the nature of the energy input (indirect/direct) plays a role. For the case of an indirect energy input, in Sect. 3.3.4 the probability q that an impinging ion subplants a surface boron atom into deeper layers (3.38) was introduced. The analogous quantity for direct energy input is the probability f that a boron ion will penetrate below the surface. From TRIM calculations, the relation

$$f = 1 - \exp\left(-\frac{E_i - E_{p'}}{E_{q'}}\right) \tag{3.52}$$

can be derived for f [72] where $E_{p'}$ and $E_{q'}$ again represent constants.

A phase transition sp^2 $\longrightarrow$ sp^3 caused by the ion bombardment requires a densification of the material already deposited. Such a densification is only possible if the flux of boron atoms able to reach positions below the surface is at least equal to the flux of atoms which are incorporated within the films[56]:

$$q \geq 1/\Phi^* = 2/F^* \qquad \text{indirect energy input} \tag{3.53}$$
$$f \geq 1/\Phi^* = 2/F^* \qquad \text{direct energy input} . \tag{3.54}$$

In Fig. 3.25, $1/q$ and $1/f$ are presented in the F^*/E diagram. According to this figure, the nucleation of c-BN is possible only under conditions which allow a densification of the material (for indirect energy input the standard region above F_h, for direct energy input arrow 1, i.e. the experiments of Hofsäss et al. [236]). This is of importance for the discussion of the nucleation mechanisms in Sect. 4.4.2. For the experiments of Hahn et al. [243] (arrow 2 in Fig. 3.25, indirect!) the condition (3.53) is not fulfilled during the growth; this means that the growth of c-BN is possible without densification. Therefore the conclusion can be drawn that the growth of c-BN takes place via attachment processes (in the sense of the preferential bond formation discussed above), in accordance with the experimentally observed crystallite sizes. Nevertheless, owing to the scarce amount of data on growth processes below F_h, a further discussion of these mechanisms would be pure speculation at the present state of knowledge.

3.3.6 Conclusions

Mechanisms of c-BN Growth. The sputter model, applied to the formulation of the competitive growth of nuclei, describes the 'standard process' of c-BN deposition even quantitatively; it cannot explain, however, the new results presented in the preceding section which present deviations from the standard process. On the other hand, sputter processes do play a relevant role during the standard process, as is evident from the low boron incorporation probabilities (Sect. 3.3.5). Also, the sputter selectivity has been proven beyond doubt. So far, at the present state of research, there is a striking analogy between the sputter model for the growth of c-BN and the model of selective chemical etching of sp² carbon by atomic hydrogen for the growth of diamond: on the one hand, selective sputtering and selective etching are obviously not the only mechanisms relevant to the corresponding process; their importance, however, on the other hand, is beyond doubt. Moreover, by means of these mechanisms the respective processes can be described rather well (in the case of the sputter model, even quantitatively); however, for both cases some important exceptions exist.

The other models which are presently controversial in the literature and which have been sketched briefly in Sect. 3.3.4 are – as the discussion in Sects. 3.3.4 and 3.3.5 has shown – even less suited to explain the existing data on the growth of c-BN (at least in the formulations published up to now). The major points of criticism can be summarized as follows (for a more detailed criticism, we refer to [97]):

- Figure 3.23 shows that the boundary of the c-BN region can be described correctly neither by the subplantation model nor by the stress models (these are the only models for which quantitative formulations exist).
- The growth of c-BN takes place via attachment of atoms to already existing crystallites; this is confirmed beyond doubt by the crystallite sizes observed, as well as by the influence of the substrate temperature. This excludes all those models postulating a phase transition within already deposited material (subplantation, stress).
- Relaxation processes always lead to the crystalline, sp^3-bonded modification, in contrast to all models predicting a relaxation towards sp^2 bonds (compare Table 3.6).

It has to be emphasized, therefore, at this point that none of the presently proposed models of ion-assisted deposition of c-BN can explain all existing data adequately.

The above discussion of the growth mechanisms of cubic boron nitride can be briefly summarized by the following points:

- The deposition of c-BN is (up to now) only possible with ion-assisted methods.
- None of the models of the deposition of c-BN published up to now is able to explain all existing data consistently. This is also true for the sputter model.
- On the other hand, the sputter model provides by far the best description of the standard process of c-BN deposition (indirect energy input; nucleation and growth take place at the same parameters). Moreover, the relevance of (selective) sputter processes has been proven beyond doubt.
- It is certain that the growth of c-BN takes place by attachment processes rather than by a phase transition of already deposited material.

Comparison with Diamond. From the above discussion it is obvious that the mechanisms of c-BN deposition are drastically different from those of low-pressure diamond synthesis. Indeed, the optimum conditions for the growth of each of the materials are rather detrimental to the growth of the other:

- Ion bombardment of diamond layers leads in most cases to their amorphization and at least to a degradation of the diamond quality (radiation damage) resulting from the formation of sp^2 bonds. This is true for high energies [246] as well as for the energy range relevant to the deposition of c-BN [247, 248].[57]
- The presence of hydrogen, which plays a decisive role during diamond deposition, is detrimental to the formation of c-BN; if the hydrogen concentration is too high, c-BN formation is suppressed even within the c-BN region (see above) [201, 202, 204][58].

First, the question arises of why radiation damage does not play any (or at least any decisive) role in the deposition of c-BN: although, on the one hand, such effects (a decrease of the c-BN content at very high ion energies) have been observed in certain experiments [210, 249], the deposition of almost pure c-BN layers is, on the other hand, possible even at ion energies as high as 1200 eV [196]. The reason is the relatively high ionicity of the BN bond (see Sect. 2.2.2). In general, the resistance of a crystalline solid against destruction (amorphization) by ion bombardment is higher the higher the ionicity of the bond involved [250]. It is therefore widely accepted that the ionicity of the BN bond plays an important role in the formation of crystallites under ion bombardment [26, 97, 218].

The purely covalent carbon–carbon bond in diamond, on the other hand, has the effect that ion bombardment easily causes amorphization of diamond or at least its graphitization. The deposition of high-quality diamond films by ion-assisted techniques has therefore proven impossible up to now [85][59].

Second, the question has to be discussed of why, up to now, all attempts to develop a process for the deposition of c-BN which is analogous to low-pressure diamond synthesis have failed. At this point, again the role of atomic hydrogen during diamond deposition has to be addressed. From the presentation in Sect. 3.2 it is evident that, of the various tasks of atomic hydrogen compiled in Table 3.2, two are of major importance: the stabilization of diamond surfaces and, closely related, the creation of growth sites by hydrogen abstraction reactions (Fig. 3.11) on the one hand and selective etching of

---

[57] Compare also Sect. 4.3.
[58] This explains also the failure of the 'chemical' CVD methods for c-BN deposition listed in Fig. 3.18, since in most cases these methods try to copy diamond deposition techniques and work with high concentrations of hydrogen.
[59] It will be shown in Sect. 4.3 that an ion bombardment at the beginning of the deposition increases the nucleation density during low-pressure diamond synthesis drastically. However, if this ion bombardment is applied during the entire deposition processes, only diamond films of very low quality are obtained.

sp^2-bonded fractions of the deposited material on the other hand. Even if, therefore, in the case of c-BN hydrogen prevents – for reasons which have not been identified yet – the formation of the tetrahedral phase, the question arises of whether these tasks can be fulfilled by other radicals (or species).

Unfortunately, there are at present no experimental papers in the literature on the nature of c-BN surfaces under those conditions which are relevant for the various deposition techniques [67]. This is especially true for the influence of hydrogen on the structure of c-BN surfaces. The situation is made even more complicated by the fact that the principal surfaces of c-BN can be either boron- or nitrogen-terminated (see Sect. 2.2.2). Nevertheless, very recently, some molecular-dynamic calculations on the nature of pure c-BN surfaces have been published [98, 251], from which it can be concluded that the {100} and {111} surfaces of c-BN, irrespective of the termination, always reconstruct in such a way that the outermost atomic layers possess an sp^2-bonded structure [98]. The growth of c-BN by attachment of species to existing crystalline faces in analogy to low-pressure diamond synthesis seems therefore to be impossible, unless a method is found to stabilize the c-BN surfaces in a way similar to the stabilization of diamond surfaces by atomic hydrogen. Since for the reasons listed above, this task cannot be fulfilled by atomic hydrogen, other radicals have to be found for this role; at the present time halogens such as F• and Cl• seem to be the most promising candidates. Concerning this aspect, however, no investigations exist in the literature to the best of the author's knowledge.

From the point of view of a possible selective (chemical) etching of unwanted sp^2 phases, the situation is similar. Apart from the experimentally observed incompatiblity of the formation of c-BN with the presence of hydrogen, it is evident from thermodynamic calculations [252] as well as from experimental investigations [52] that a significant selectivity of the etch rates of h-BN and c-BN can be obtained neither with molecular nor with atomic hydrogen; Harris et al. [253] even report that under the conditions of diamond synthesis (hot filament), hexagonal boron nitride is etched significantly neither by molecular nor by atomic hydrogen[60]. Thus in this case, also other species have to be used; and again, halogens seem to be the most promising candidates. In some of our own experiments, no selectivity with respect to the bond structure of BN films has been found with reactive ion etching with either BCl_3 or SF_6 [72]. Calculations of Bohr et al. [52] show that under certain conditions c-BN is more stable than h-BN in the presence of chlorine compounds; however, these predictions still have to be confirmed experimentally. Fluorine-containing compounds, on the other hand, seem to be not suited to this task according to the analyses of Bohr et al. A new publication of Schaffnit et al. [256], however, seems to indicate that h-BN

---

[60] By the addition of 1 % CH_4, i.e. under the conditions of diamond growth, a selective etching of h-BN seems to be possible – with very low etch rates, however – [253, 254] (it may be mentioned that CH_4/H_2 mixtures are used to etch III/V semiconductors such as InP [255]).

is etched selectively by an Ar/Cl$_2$ plasma. Summarizing these observations and calculations, it has to be stated that the existing data seem not to be sufficient to decide whether a selective etching of h-BN with respect to c-BN can be obtained by the use of halogen compounds[61].

Therefore, at this point the question remains unsolved as to whether – in analogy to the low-pressure synthesis of diamond – a 'chemical route' to c-BN deposition is possible, i.e. a deposition process without strong ion bombardment. In order to answer this question, basic investigations concerning the nature of c-BN surfaces and possible methods for their stabilization by various radicals are required, and also experiments concerning the selective etching of h-BN and c-BN.

Despite all these differences, however, there is one most important aspect which is common to the low-pressure synthesis of diamond and the ion-assisted deposition of c-BN: in both cases, the growth takes place by attachment processes to already existing crystallites, the formation of which (i.e. nucleation), however, has to be discussed separately in each case (Sects. 4.2 and 4.4, respectively). Both growth by attachment processes and a separate nucleation are not true for tetrahedral amorphous carbon.

3.4 Growth of ta-C Films

Although in our own group no experimental investigations concerning the deposition of tetrahedral amorphous carbon have been carried out, it seemed nevertheless necessary to study intensively the fundamentals of the deposition of ta-C for two reasons. On the one hand, most of the models of the deposition of c-BN briefly presented above were originally developed for ta-C (as already mentioned). In order to judge these models it was therefore necessary to investigate in some detail the mechanisms of ta-C deposition. On the other hand, the ion-assisted deposition of a mainly sp^3-bonded carbon modification is concerned here. The deposition of ta-C forms, therefore, an important connecting link between c-BN deposition on the one hand and low-pressure diamond synthesis on the other hand; an intensive investigation of the elementary mechanisms of this process is therefore of major importance for the main topic of this book, the evaluation of possible common aspects of the deposition of diamond-like superhard materials.

---

[61] According to the molecular-dynamic calculations mentioned above [98, 251], c-BN surfaces always reconstruct towards sp^2-BN, even if the extent of the reconstruction and the underlying mechanisms are different for the various principal surfaces. This would mean that a selective etching of h-BN with respect to c-BN is impossible since the first one or two monolayers of a BN film alway show sp^2 character. On the other hand, selective physical etching (sputtering) can play a role despite this reconstruction, since sputter processes can also take place below the surface down to the penetration depth of the ions. In order for selective chemical etch processes to become effective, a stabilization of the existing c-BN surface is (as in the case of diamond) a sine qua non.

3.4.1 Experimental Observations

We shall start again with a short description of the techniques which can be used for the deposition of ta-C. It has already been mentioned that the formation of ta-C also requires the bombardment of the growing film with ions. Depending on the nature of the energy input, the methods can be divided into two groups [215]:

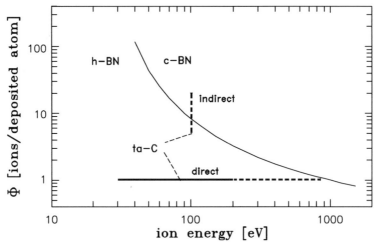

Fig. 3.27. Regions of successful ta-C deposition in the Φ^*/E_i diagram. For comparison, the boundary between the h-BN and c-BN regions is shown also [72]

- In contrast to c-BN deposition, techniques are usually applied that work with direct energy input, i.e. with energetic carbon ions. The ratio of ions to film-forming species Φ is therefore unity. Among the methods most often employed are ion beam deposition (IBD) [257, 258, 259, 260] and vacuum arc deposition [261, 262, 263, 264, 265, 266]. In general, ta-C deposition is performed (at least nominally) at room temperature. By means of plasma beam deposition, the formation of ta-C:H films is also possible [92] (see Sect. 2.4.2).
- Some few papers also exist on deposition with indirect energy input, where carbon is provided with low energies only, while simultaneously the growing film is bombarded with inert-gas ions (e.g. magnetron sputtering [267]). For these processes, rather high ions/neutrals ratios are required ($\Phi \approx 5$, see Fig. 3.27) [267]; the formation of ta-C with $\Phi \approx 1$, as in the case of direct energy input, is not possible [268].

Now, the most important parameter dependences will be described briefly (Figs. 3.28 and 3.29); here the comparison with c-BN always forms an essential aspect.

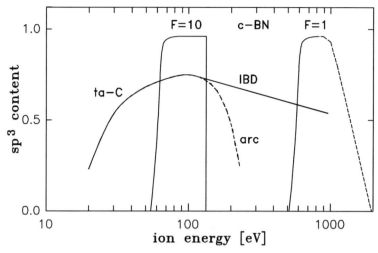

Fig. 3.28. Comparison of the energy dependences of the sp^3 content of c-BN and ta-C, (schematic). In the case of c-BN, curves are given for $F = 1$ and $F = 10$

The dependence of the sp^3 content q_t of ta-C films on the ion energy (for the case of direct energy input) is presented in Fig. 3.28; q_t first increases gradually with increasing ion energy, reaching a maximum at about 100 eV. If the ion energy is enhanced further, q_t starts to decrease; however, differences exist for the various deposition techniques: in the case of ion beam deposition, the decrease is rather gradual; it is possible to obtain films with $q_t > 50\%$ even at 1000 eV [259, 260]. For the vacuum arc technique, on the other hand, the sp^3 content drops rather sharply with E_i [264, 266]. These differences for the various methods can presumably be explained by a secondary effect [72, 215]: in the case of arc deposition, the high ion currents inherent in this technique lead, at high energies, to a strong heating of the substrate. High temperatures, however, are detrimental to the formation of sp^3 bonds (see below); it is therefore reasonable to assume that the energy dependence of q_t is represented by the curve for the IBD method in Fig. 3.28.

Figure 3.28 also shows for comparison the energy dependence of the sp^3 content q_c of c-BN films. Here it has to be taken into account that the energetic position of the c-BN region depends on the ion/neutral ratio F (Fig. 3.20)[62]. Therefore, in Fig. 3.28, $q_c(E_i)$ curves are presented for $F = 1$ and $F = 10$.

Besides this dependence of $q_c(E_i)$ on F, the major difference between ta-C and c-BN can be found in the sharpness of the transitions: whereas q_t increases only gradually with E_i and decreases even more slowly, the boundaries of the c-BN region with respect to the ion energy are very abrupt: by

---

[62] In contrast, for the case of ta-C only $\Phi = 1$ is considered in Fig. 3.28, i.e. direct energy input.

an increase of the energy of about 10%, the c-BN content can be switched from zero to almost 100%. On the other hand, the deposition of c-BN takes place near the resputter limit; this explains the sharp rupture of the c-BN region on the high-energy side. The deposition of ta-C is, in contrast, carried out far from the resputter limit.

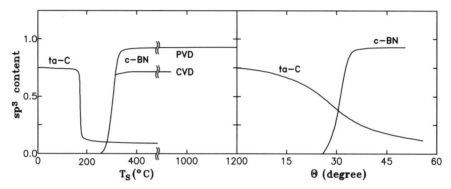

Fig. 3.29. Comparison of the dependence of the sp^3 content on the substrate temperature and the ion angle of incidence for c-BN and ta-C (schematic)

Figure 3.29 shows the dependence of q_t and q_c on the substrate temperature T_s and the ion angle of incidence Θ_i. In both cases, the trends are opposite for ta-C and c-BN. High temperatures as well as high angles of incidence favor the growth of c-BN; in principle it is possible to switch from the h-BN to the c-BN region by an increase of either of these parameters. On the other hand, either large Θ_i [263, 269] or a high temperature [257, 260, 266, 270] is detrimental to the formation of ta-C. The influence of T_s is especially pronounced: Fig. 3.29 shows a rapid decline of q_t at about 200°C.

Finally, some properties of ta-C films are listed which are important for the following discussion: the sp^3 content of ta-C films seems to be limited to about 80% whereas c-BN layers – if the nucleation layer (see Sect. 4.4) is neglected – can be deposited with a cubic fraction of almost 100%. Like c-BN layers, ta-C films possess a high compressive stress, in the GPa region. On annealing of ta-C layers, a retransformation towards sp^2 material is observed above 450–750°C [265, 271, 272], whereas c-BN is thermally stable up to at least 900°C [273, 274, 275].

3.4.2 Growth Mechanisms

The above presentation of the main techniques for the deposition of ta-C and, especially, the relevant parameter dependences reveal that, although the growth of both ta-C and c-BN is only possible with ion bombardment and takes place in a similar parameter space, essential differences exist between the two processes. It is therefore evident that the growth of ta-C and c-BN cannot – as often proposed [30, 50, 230] – rely on the same mechanisms.

The models of the deposition of ta-C published in the literature are very similar to the models of the growth of c-BN which have been discussed above in Sect. 3.3.4. Again a classification into two groups of descriptions is necessary [215], which are distinguished mainly by the role thermal spikes play during the formation of ta-C (Table 3.6).

The *quenching model* assumes, as in the case of c-BN (Sect. 3.3.4), that thermal spikes lead, according to (3.36), to HPHT-like conditions which favor the formation of sp^3 bonds. It has to be taken into account, however, that under these HPHT conditions diamond is the stable phase of carbon whereas ta-C is still metastable. Therefore the high quenching rates of 10^{14}–10^{15} K/s are thought to be responsible for the fact that the stable crystalline phase is not obtained but, rather, an amorphous, mainly sp^3-bonded network is frozen-in. Molecular-dynamic simulations of such quenching processes [276, 277] indeed result in amorphous structures with the properties observed experimentally.

The second group of models is, in contrast, based on the assumption that the sp^3 bonds are already formed during the collision cascade (Fig. 3.19). This leads to the generation of defects, mainly interstitials, which in turn cause either a local densification (*subplantation model*) [225] or the formation of compressive stress (*static stress model* [226, 261, 278]); both effects lead to the formation of sp^3 bonds according to the mechanisms already discussed in Sect. 3.3.4. The thermal spike following the collision cascade, on the other hand, induces relaxation processes according to (3.37). A quantitative description is therefore again obtained by means of a balance equation of defect generation by the ion bombardment on the one hand, and defect relaxation by the spikes on the other hand. In the case of the stress model, (3.40) again follows. In the subplantation model, for the densification one has to take into account that in the case of ta-C the energy input is in general direct; therefore the penetration probability f of an ion, given by (3.52), is decisive; thus for the densification it follows that

$$\rho = \rho_0 \cdot \left(1 + \frac{f}{\frac{1}{\Phi^*} - f + 0.0016p \left(\frac{E_i}{E_0}\right)^{5/3}} \right) , \qquad (3.55)$$

in analogy to (3.39) for the case of indirect energy input.

The energy dependence of ta-C growth presented in Fig. 3.28 for the vacuum arc deposition method can be fitted, with a suitable choice of parameters, by (3.40) (static stress model) as well as by (3.55) (direct subplantation) [215]. This, however, is not true for the more general form of the energy dependence for ion beam deposition. This deviation could be caused by an inadequate

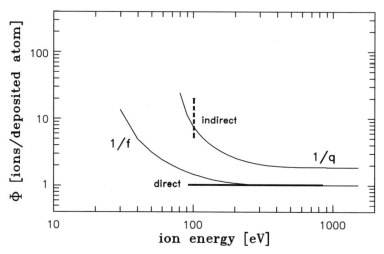

Fig. 3.30. Comparison of the parameter regions for the deposition of ta-C with direct and indirect energy input with the probabilities of direct and indirect subplantation ((3.52) and (3.38)) in the Φ^*/E_i diagram

description of the relaxation processes during the thermal spikes in the approach of Seitz and Koehler (3.36), the application of which to the deposition of c-BN and ta-C films has recently been criticized by several authors[63].

Figure 3.27 shows that in the case of ta-C, differences also exist with respect to the form of the energy input. These are taken into account by the subplantation model but not by the stress model. From Fig. 3.30 it is evident that during the deposition of ta-C the conditions for effective densification (3.54) and (3.53) are fulfilled for direct as well as for indirect energy input. Summarizing, the above discussion yields the result that the growth of ta-C relies on local penetration processes, which lead to a densification of the deposited material, which in turn causes the generation of sp^3 bonds. Thus, the formation of tetrahedral amorphous carbon takes place in the interior of the films by local processes. It can, at least qualitatively, be described by means of the subplantation model, which nevertheless still has to be improved to yield a quantitative description also.

According to the model of random covalent networks of Angus and Jansen [117], ta-C layers are overconstrained (see Sect. 2.4.2). Relaxation process, caused either by thermal spikes or by high substrate temperatures, therefore lead always towards sp^2 bonds. The formation of ta-C has therefore to be

---

[63] Equation (3.36) is based on the assumption of a spherical spike. During the Fourth German 'c-BN expert meeting' in Reichenau, June 1997, the discussion led rather unanimously to the view that for the deposition processes under consideration cylindrical regions of high phonon density have to be taken into account. A quantitative analysis has been published very recently by Hofsäss et al. [227]; however, this important paper still awaits thorough discussion in the literature.

regarded as a 'freeze-in' process; the layers obtained are not thermally stable. This explains the temperature dependence of the ta-C content (Fig. 3.29) as well as the influence of annealing processes.

3.4.3 Conclusions

Finally, the deposition of ta-C will be briefly compared with the growth of c-BN (ion-assisted deposition), on the one hand, and the low-pressure synthesis of diamond (also an sp^3-bonded modification of carbon), on the other hand.

Comparison of Diamond and ta-C. Tetrahedral amorphous carbon is an amorphous sp^3 carbon modification, the deposition of which relies on densification processes which are realized by ion bombardment. During low-pressure diamond synthesis, in contrast, no postdeposition phase transformation occurs; rather, the growth takes place by attachment of species to already existing crystallites or nuclei. Thus, the two processes are completely different; as for the comparison of diamond and c-BN (Sect. 3.3.6), the respective deposition conditions exclude each other: diamond will undergo graphitization or amorphization under ion bombardment; however, owing to the high temperatures, not ta-C but, rather, sp^2 carbon is formed. On the other hand, atomic hydrogen, which is present in high concentrations during diamond synthesis, etches amorphous carbon with much higher rates than diamond.

Comparison of c-BN and ta-C. The presentation of the major parameter dependences of ta-C deposition in Sect. 3.4.1, and the discussion of the mechanisms underlying these dependences in Sect. 3.4.2 have shown that there are serious differences between the deposition of ta-C and that of c-BN, although each of the materials is formed under strong ion bombardment in a very similar parameter space. The main differences can be briefly summarized as follows:

- In the case of ta-C the phase formation relies on a transformation $sp^2 \longrightarrow sp^3$ of already deposited material caused by the ion bombardment, whereas the deposition of c-BN takes place by attachment processes to existing nuclei or crystallites.
- Relaxation processes, which may be induced either by high substrate temperatures (or by post-deposition annealing) or, especially, by the high local temperatures generated by the thermal spikes, favor the formation of sp^3 bonds in the case of c-BN and lead thus to an improvement of the quality of the c-BN layers[64]; in the case of ta-C, however, they favor, in sharp contrast, the (re)formation of sp^2 bonds.

---

[64] This becomes evident in, for example, the recombination of defects, which leads to, among other things, a stress reduction at high substrate temperatures or after post-deposition annealing [214, 240, 273, 275, 279, 280]; in addition, as has been discussed in Sect. 3.3.1, a minimum temperature of about 150°C exists for the deposition of c-BN whereas there is, in contrast, a maximum temperature for the deposition of ta-C (Fig. 3.29).

112   3. Growth Mechanisms

Before debating possible reasons for these differences in detail, it seems to be appropriate to enlarge the underlying set of data once more by considering another, fourth material, i.e. β-C_3N_4; the present discussion will then be continued in Sect. 3.6.

3.5 Growth of CN Films

3.5.1 Experimental Observations

In the paper in which Liu and Cohen proposed β-C_3N_4 as a candidate for possible superhard carbon nitride compounds, these authors suggest – in analogy to the synthesis of diamond and c-BN – the use of HPHT techniques also for the preparation of β-C_3N_4, i.e. the synthesis of bulk material [22]. However, the number of articles in the literature on HPHT synthesis of β-C_3N_4 is small [281, 282, 283] (see also Table 3.7); all attempts reported up to now have been without success. On the other hand, there is already a considerable number of papers on experiments to prepare β-C_3N_4 by thin-film technology methods – similarly, almost always unsuccessful. Interestingly, most of the attempts do not try to copy the conditions of either diamond or c-BN deposition[65]. The most important of these techniques are summarized in Table 3.7; they will be discussed in the following in a brief survey[66].

In quite a number of the papers listed in Table 3.7, and in some further papers published in the literature, a successful deposition of β-C_3N_4 has been postulated without, however, providing real evidence[67]. Only in a limited number of experiments have nanocrystallites been obtained with diffraction patterns which agree roughly with the predicted lattice parameters of β-C_3N_4 [290, 295]. Nevertheless, in these cases also, a final proof has still to be provided[68]. The most important question arising at this point addresses therefore the reasons for the failure of (almost) all past attempts to synthesize β-C_3N_4.

In Table 3.7, in addition to the methods applied and the temperatures and ion energies used in these methods, some data concerning the stoichiometry and the structure of the films obtained with these techniques are also listed. This table already illustrates the three major problems which have to be solved in order to accomplish a successful deposition of β-C_3N_4:

---

[65] Rather, parallels exist to the deposition of ta–C.
[66] Up to now, there have been only a few review articles on the deposition of β-C_3N_4; some important aspects have been summarized by Marton et al. [71] and, very recently, in [284, 285].
[67] Especially striking examples of obviously wrong interpretations can be found in [291, 298].
[68] These uncertainties and also the indeed sometimes striking misinterpretations result of course, to a large extent, from the problems inherent in the characterization of β-C_3N_4 as discussed in Sect. 2.4.4.

Table 3.7. Compilation of methods for the deposition of CN films. This table does not in any way claim completeness; further publications on the deposition of β-C_3N_4 can be found in, for example, [71]. For all methods, the ranges of ion energies and substrate temperatures applied are given. 'RT' means room temperature, and 'ns' that the respective parameter has not been specified. The final columns address properties of the deposited films

Author	Method	E_i (eV)	T_s (°C)	x^a (%)	Structure[b]	sp^3 [c]	IR[d]
HPHT conditions							
Wixom [281]	Shock wave	—	ns	—	Diamond + N_2		
Sekine [282]	HPHT pyrolysis	—	1400	18	Crys.	o	—
Diamond conditions							
Badzian [125]	MWCVD	—	1000	<0.5	Diamond		
c-BN and ta-C conditions							
Fujimoto [286]	IBAD (e-beam)	10k, 20k	≤100	50[e]	a	o	—
Fujimoto [286]	IBAD (e-beam)	200, 500	≤100	70[e]	a	o	—
Hofsaess [287]	MSIBD	20–500	RT, 350	30	a	—	● ◊
Rossi [288]	IBAD (sputtering)	50–350	ns	32	a	—	● –
Chen [289]	DC sputtering	0–200	ns	42	nc[f], a	o[g]	● ◊
Sjoestroem [102]	DC sputtering	0–200	150–600	28	t	●	● ◊
Yu [290]	RF sputtering	ns	400–600	50	nc, a	—	—
Weber [291]	Plasma beam	100	ns	40	a	—	● ◊
Cuomo [292]	RF sputtering	50	20–450	50	a	—	– ◊
Marton [293]	MSIBD	5–50	ns	40	a	●	—
Bousetta [294]	ECR, e-beam	20–30	100–700	45	a	●	◊
Further deposition techniques							
Veprek [104]	PECVD	≪500[h]	800	69	a	—	●
Niu [295]	Laser ablation	—	160–600	47	nc, a	—	● ◊
Zhao [296]	Laser ablation	—	ns	32	a	—	● ◊
Bulir [101]	Laser ablation	—	RT	30	a	—	● ◊
Ion implantation							
Hoffman [297]	Implantation	500	RT–500	14	a	●	—
Marton [293]	Implantation	5–75	ns	40	a	●	—

[a] Maximum nitrogen content of the $C_{1-x}N_x$ film. [b] a, amorphous; t, turbostratic; nc, nanocrystalline. [c] ●, according to the analysis of Marton et al. [71, 100], part of the carbon is bonded tetrahedrally; o, the carbon is exclusively sp^2-bonded; —, no measurements have been presented. [d] ●, existence of the cyanide band (C≡N) at 2100–2200 cm^{-1}; ◊, existence of typical graphite bands between 1300 and 1600 cm^{-1} in IR and Raman spectra; —, no IR or Raman spectra available. [e] These values are, as has also been pointed out by [71], quite unrealistic. [f] No details of the diffraction pattern reported. [g] Concluded from EELS measurements. [h] In the paper, the bias voltage is given. Owing to the high working pressure (0.3 mbar), the ion energy is much lower.

1. Almost all of the $C_{1-x}N_x$ films obtained up to now are markedly understoichiometric.
2. The films are, in almost all cases, amorphous; only in a few cases have nanocrystalline particles embedded in an amorphous matrix been observed (see above).
3. According to IR or Raman spectra, most of the films contain C≡N groups as well as sp^2-bonded, graphite-like carbon, thus deviating strongly from the desired bond structure in which the carbon is bonded tetrahedrally and every nitrogen atom forms bonds to three C atoms.

In the following, possible reasons for the missing stoichiometry, the missing crystallinity and the appearance of unwanted bond structures will be discussed by means of a brief survey of the most important experiments in the literature.

Stoichiometry. Carbon nitride films with a composition corresponding to the stoichiometry of β-C_3N_4 have been well-known for many years [71, 104]. However, these are polymer-like films with properties which are completely different from those of a crystalline, diamond-like superhard material. If, during the deposition, bombardment with energetic particles is applied, which up to now has been the case in most attempts to deposit β-C_3N_4, the maximum nitrogen content reached is typically in the region between 30 and 40 % (Table 3.7). In order to explain this understoichiometry, the following experimental observations are of importance:

1. the nitrogen content x of the films first increases with increasing nitrogen supply but gradually reaches a saturation value [71, 101, 287, 296];
2. x decreases with increasing ion energy and increasing ion flux [287, 299, 300];
3. x decreases with rising substrate temperature [102, 105, 287, 300];
4. especially high nitrogen concentrations are obtained in those cases in which the nitrogen is supplied in atomic form [104, 295].

These facts are strong indications that the understoichiometry is mainly caused by two effects: in the first instance, molecular nitrogen in the gas phase is relatively stable; thus, those techniques are to be preferred in which nitrogen is supplied in atomic or ionic form. On the other hand, Boyd et al. [293] reach, by means of MSIBD, despite a ratio of primary ions of $N^+/C^+ = 8{:}3$, a maximum nitrogen concentration of only 40%. It is now generally agreed (e.g. [26, 71, 299]) that chemically enhanced preferential sputtering of nitrogen with respect to carbon[69] is responsible for this limitation. The dimeric molecules N_2 and CN, as well as C_2N_2, are extremely volatile; once formed – for example by the influence of ion bombardment – they will immediately leave the growing film. The formation of such dimers, however, becomes more

---

[69] This effect is also confirmed, for example, by AES measurements by Marton et al. [71] and by XPS analyses by Bulir et al. [101].

probable the higher the nitrogen concentration already is. This explains the saturation effect with increasing nitrogen supply [71]. Likewise, chemically enhanced preferential sputtering of nitrogen is the reason for the decrease of x with increasing ion bombardment and increasing substrate temperature.

Crystallinity. The scarce data published up to now in the literature on nanocrystalline $C_{1-x}N_x$ particles (see Table 3.7) are not sufficient by far even to speculate on possible parameter dependences. Therefore, at this point, only the important fact should be stressed that the ionicity of the carbon–nitrogen bond in β-C_3N_4 is very low [31] (see also Table 3.8). Thus, a very important driving force towards crystallinity is missing which – as has been pointed out several times – plays an essential role in the case of c-BN; it is therefore to be expected that even a relatively weak ion bombardment leads to amorphization. This aspect will be discussed more comprehensively in the following sections, especially in Sect. 3.6.

Bond Structure. In almost all IR and Raman spectra of $C_{1-x}N_x$ films the peak of the cyanide group (C≡N) and bands corresponding to those of graphite or sp^2-bonded amorphous carbon are observed (see Table 3.7)[70]. The existence of the cyanide group shows that the nitrogen is at least in part bonded to only one carbon atom, whereas bonds to three C atoms are required for β-C_3N_4; the graphite-like bands indicate that the carbon is sp^2-bonded rather than forming tetrahedral bonds. From these IR and Raman spectra it is therefore obvious that most of the films obtained up to now – apart from the missing crystallinity – do not possess the bonding structure required for β-C_3N_4. This is also comfirmed by the density of these films, which in most cases is only about $2.2\,\text{g cm}^{-3}$ [6, 287, 288], which is much less than the theoretically predicted value of $3.54\,\text{g cm}^{-3}$ but which is very similar to the densities of graphite and h-BN (Table 2.1).

However, according to an analysis by Marton et al. [71, 100], XPS spectra indicate that in some cases the amophous $C_{1-x}N_x$ films consist of two phases, one tetrahedrally bonded and one trigonal[71]. In this case, the tetrahedral phase possesses the stoichiometry required for β-C_3N_4, whereas the composition of the sp^2 phase varies but is always understoichiometric.

It can be seen from Table 3.7 that – as far as the existing data allow one to draw a conclusion – those films containing this tetrahedral phase (reserving a final confirmation of this identification) are obtained especially at low energies, in agreement with the observation of Marton et al. that the concentration of this sp^3 phase decreases for energies above 40 eV.

Molecular-Dynamic Calculations. In the context of the principal question of whether the deposition of β-C_3N_4 or other high-density CN modifications is possible at all, new molecular-dynamic calculations by Frauenheim's

---

[70] The cyanide group is especially found in films with a high nitrogen content [296, 301].
[71] A confirmation of the identification of these phases by other characterization techniques (e.g. EELS) has yet to be performed.

group yield some interesting insights [302]. Although these investigations relate to amorphous carbon nitride, they nevertheless provide a number of very important facts.

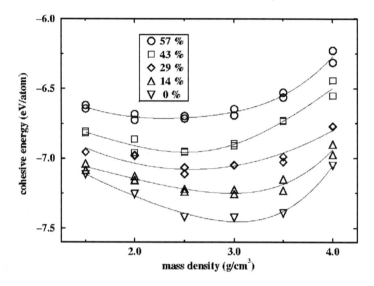

Fig. 3.31. Results of molecular-dynamic calculations: bond energy of amorphous carbon nitride films of different compositions as a function of the density [302]

The perhaps most important result is presented in Fig. 3.31, which shows the binding energy as a function of the density for carbon nitride films of various compositions. For $x = 0$, the energy minimum is found slightly above $3 \mathrm{\,g\,cm^{-3}}$, in agreement with the density typically found for ta-C films. But from Fig. 3.31 it can be seen that the energy minimum shifts steadily to lower densities as the nitrogen fraction of the material becomes higher. For $x = 0.57$, i.e. the stoichiometry of β-C_3N_4 films, E_c is situated at about 2.2 $\mathrm{g\,cm^{-3}}$, i.e. the density of graphitic films.

Further calculations show that, at constant density, the fraction of sp^3-bonded carbon decreases drastically if the nitrogen content is increased. Dense structures with a high content of nitrogen possess, rather, a low degree of three-dimensional crosslinking; they contain chain- and sheet-like structures and an increasing fraction of double and even triple bonds.

These results of molecular-dynamic calculations are strong indications that the synthesis of carbon nitride films with a high content of nitrogen, a high density and a large fraction of tetrahedrally bonded carbon is difficult or maybe even inpossible.

3.5.2 Discussion

The deposition of large-area crystalline β-C_3N_4 layers, analogous to the state of the art in the cases of polycrystalline diamond films and nanocrystalline c-BN films, has definitely not been achieved up to now. Nevertheless, it cannot be excluded that in quite a few cases isolated β-C_3N_4 nanocrystallites have been obtained; moreover, under certain conditions the deposition of amorphous $C_{1-x}N_x$ with a distinct fraction of sp^3-bonded carbon seems to be possible (see Table 3.7). However, even if the positive identifications of β-C_3N_4 material claimed in the literature are accepted, the existing data are too incomplete and too unsystematic to allow one to discuss at the present time possible growth mechanisms or even to quantify parameter ranges for the growth of crystalline or amorphous tetrahedral carbon nitride films.

Therefore at the present time – especially in view of the main topic of this book – the question of interest is, rather, why the special conditions which can be applied successfully to deposit other superhard materials (diamond, c-BN, ta-C) obviously do not lead to success in the case of carbon nitride.

All *HPHT experiments* [281, 282, 283] performed up to now in analogy to diamond and c-BN synthesis result either in a complete separation into crystalline carbon and N_2 [281, 283] or at least in a drastic loss of nitrogen [282]. The remaining carbon is present either as graphite or – if extremely high pressures are applied [281][72] – as diamond.

In this context, some calculations of Liu and Wentzcovich are worth mentioning which show that under these conditions the system (diamond + N_2) is energetically more favorable than the various hypothetical tetrahedral or graphitic carbon nitride modifications under discussion [32].

To the best of the author's knowledge, no papers exist in the literature on the deposition of carbon nitride films under the *conditions of low-pressure diamond synthesis* (high gas phase temperatures, substrate temperatures between 700 and 1000°C, hydrogen-rich atmosphere). On the other hand, there are a number of publication on the synthesis of diamond in the presence of nitrogen in the gas phase. Badzian et al. [125] show that MWCVD deposition of diamond (even if of poor quality) is possible from CH_4/N_2 mixtures. However, within the detection limits of the characterization techniques applied ($\leq 0.5\%$), no nitrogen is incorporated into the diamond lattice.

In addition there are several papers on doping of diamond films with nitrogen during either HFCVD or MWCVD deposition [51, 187, 303]. All these processes have, however, in common the fact that, despite sometimes rather high nitrogen concentrations in the gas phase [51], the resulting nitrogen content of the layers is extremely low (some hundred ppm maximum)[73].

---

[72] In the shock wave experiments of Wixom et al., pressures in the range of 60–250 GPa have been used [281].

[73] On the other hand, the simultenous incorporation of nitrogen and boron seems to be possible with larger concentrations [304].

These results show that – under the special condition of low-pressure diamond synthesis – nitrogen is incorporated into the diamond lattice in very low concentrations only. As is evident from the experiments of Badzian et al., this is true even for extremely nitrogen-rich and hydrogen-poor atmospheres. The underlying reasons are not understood at all; however, it is imaginable that here also the fact that the system (diamond + N_2) is energetically more favorable than crystalline carbon nitride films plays a role[74].

In the context of these attempts to deposit carbon nitride either under HPHT conditions or under the conditions of low-pressure diamond synthesis, it is also worth mentioning that in natural diamonds[75] nitrogen is not incorporated on lattice sites but, rather, forms vacancy complexes and agglomerates [187, 305, 306]. In these Ia diamonds the nitrogen concentration can reach up to 1000 ppm, whereas in synthetic IB diamond, where the nitrogen is incorporated substitutionally, the maximum content is limited to 500 ppm [187]. These facts also indicate that it is difficult to integrate nitrogen into the diamond lattice.

Since under the special conditions of diamond synthesis nitrogen is hardly incorporated into the diamond lattice at all, growth of β-C_3N_4 by attachment of well-defined species to existing crystals seems to be not possible. In order to prepare a tetrahedral phase, therefore, a densification process is required, e.g. by *ion-assisted deposition* in analogy to the cases of c-BN and ta-C. Ion bombardment, however, leads – as we have seen above – to a drastic loss of nitrogen by chemically enhanced sputtering; as a consequence, at the present time it seems to be impossible to reach the stoichiometry necessary for β-C_3N_4 by techniques working with strong ion bombardment. For the same reason, the substrate temperature cannot be chosen unlimitedly high in order to support crystallinity. Finally, owing to the low ionicity of the C–N bond, it seems improbable that ion- or temperature-induced relaxation processes can assist the formation of crystallites to the same extent as is the case for c-BN.

It is evident from the above discussion that up to now it has not been possible to transfer any of those methods which in the past have been successfully applied for the deposition of the other diamond-like materials to the case of β-C_3N_4; the existing results seem even to indicate that these methods are not suited to the deposition of β-C_3N_4 for physical reasons. Rather, according to the present state of knowledge, it has to be concluded that in order to synthesize β-C_3N_4 or any other superhard carbon nitride modification, it will probably be necessary to find some completely new approach. Indeed, the most interesting results up to now have been obtained with techniques which cannot be classified in any of the above categories (see Table 3.7):

---

[74] Jin and Moustakas [187] postulate that the low binding energy of the N–C bond is responsible for the low incorporation probability; as a consequence, enhanced etching of nitrogen by atomic hydrogen takes place under the conditions of diamond deposition.

[75] Nitrogen is the major contaminant in type I diamonds.

- Veprek et al. [104] use an inductively coupled nitrogen plasma of high plasma density in order to deposit amorphous carbon nitride films from a graphite target located inside the plasma by means of chemical transport reactions (CTRs), i.e. the formation of CN and C_2N_2 species[76]. To the best of the author's knowledge this is, up to now, the only technique by means of which the stoichiometry required for β-C_3N_4 (57 % nitrogen) can easily be reached. IR spectra indicate a still-unidentified CN modification which is clearly distinguishable from the graphite-like structures which are typically deposited by ion-assisted techniques.
- Niu et al. [295, 309] use laser ablation of a graphite target while simultaneously subjecting the growing film to a beam of nitrogen *atoms*. This leads to a relatively high nitrogen content even if the stoichiometry of β-C_3N_4 is not reached (Table 3.7); in addition, the layers contain nanocrystallites with sizes up to 10 nm; however, identification of these as β-C_3N_4 has several times been challenged in the literature [27, 71, 310, 299].
- The 'pulsed-laser-induced reactive quenching' which has been proposed by Sharma et al. [311] is a rather unusual technique. Here, the interface of an organic liquid with a solid surface is subjected to laser pulses (248 nm, 5 Hz pulses of 20 ns, 3.5 J/cm^2). The liquid source compounds are selected according to stereochemical reasons. According to the authors, cyclohexane, for example, can be used for the synthesis of diamond. For the preparation of carbon nitride, hexamethylenetetramine ($C_6H_{12}N_4$) and (for reasons of stoichiometry) liquid ammonia were chosen; a tungsten foil kept at $-60°C$ was used as the substrate. From electron diffraction patterns, the authors conclude the presence of α-C_3N_4 and β-C_3N_4; further data concerning the properties of the material, especially its stoichiometry, have nevertheless not been presented up to now.

This latter publication has been mentioned not as an example of the successful deposition of β-C_3N_4 but, rather, to emphasize the necessity to develop new techniques for the synthesis of β-C_3N_4, since the techniques developed for the other diamond-like materials can evidently not be applied in this case. To what extent methods such as that proposed by Sharma et al. will indeed bring about success is open to future investigations.

These few results are of course not sufficient as a starting point for a more detailed discussion, the more so since in all cases a final identification of the material obtained yet has to be performed. Therefore, at the present point of discussion we can only state that the deposition of the superhard material β-C_3N_4 has not been accomplished up to now, and that in all probability

---

[76] Standard PECVD methods with gaseous precursors cannot be used in the case of β-C_3N_4, since these precursors in almost all cases contain hydrogen. The incorporation of hydrogen, however, always leads to soft, polymer-like films [307]. In order to establish a PECVD method for β-C_3N_4 deposition, at least the development of new precursors is necessary [308].

the well-known techniques of diamond, c-BN and ta-C deposition cannot be transferred to this case.

3.6 Conclusions

The discussion in the first two chapters of this book raised the question of whether the low-pressure synthesis of diamond is simply a peculiarity of the (hydro) carbon chemistry, or whether the mechanisms of diamond synthesis could be applied in order to prepare other materials from the B/C/N system thus allowing the deposition of superhard materials with unique and interesting properties (Sect. 2.2.2).

On the basis on the results and analyses presented in the preceding sections, at the present time this question has to be answered by saying that low-pressure diamond synthesis is indeed a special process which is not transferable. This exceptional position is caused by atomic hydrogen which performs several most important tasks during diamond deposition (Sect. 3.2.1).

In sharp contrast, hydrogen is, especially in radical form, detrimental to the deposition of other superhard materials. During the deposition of c-BN it prevents – for reasons which still have to be identified – the formation of the sp^3 modification; in the cases of ta-C and $C_{1-x}N_x$ there is always the risk of the incorporation of hydrogen, which would lead to films with very low densities.

In order to transfer the mechanisms of diamond deposition to other materials, the role of hydrogen has to be taken over by other radicals. Oxygen-containing species such as $OH^\bullet$, which during diamond growth can at least in part fulfill the tasks of $H^\bullet$, have to be excluded, at least in the cases of boron-containing materials as there is always the risk of the formation of boric acid $B(OH)_3$ and boron oxide B_2O_3. This limits the choice to halogen radicals such as $F^\bullet$ and $Cl^\bullet$. It has indeed been shown that halogens can replace atomic hydrogen during diamond deposition, at least in part [127, 128, 129][77]. The underlying mechanisms are nevertheless still unknown to a large extent.

In the case of c-BN, chemical deposition by utilization of a halogen chemistry has been proposed several times [52, 252, 312]. Experimental investigations in order to achieve this aim, however, have been rather scarce up to now. In addition, the few studies published up to now [52, 252, 256, 312] concentrate almost exclusively on the question of selective chemical etching of h-BN with respect to c-BN by means of halogen-containing compounds. There is no doubt that selective etching indeed plays an important role during low-pressure diamond synthesis; but it is also beyond doubt that the mechanisms of surface stabilization, creation of free growth sites and hydrogen abstraction from surface adsorbates are far more essential for the deposition

---

[77] This aspect has been omitted in the above presentation of diamond growth for reasons of space.

3.6 Conclusions 121

of large, well-faceted crystals. Unfortunately, at the present time almost no studies exist on the influence of hydrogen and halogen radicals on the structure and properties of the various surfaces of c-BN and other materials in the B/C/N system [98]. It is, therefore, at the present stage of knowledge, not possible to judge definitively whether a chemical route for the deposition of cubic boron nitride (and other diamond-like materials) exists. It is, however, certain that such a route has not been found up to now.

At the present state of technology, the most important difference between diamond, on the one hand, and the other superhard materials in the B/C/N system, on the other hand, can be found in the fact that diamond – although at relatively high temperatures – grows 'voluntarily'. For all other materials, a 'driving force' in the form of an ion bombardment seems to be necessary; nevertheless – again according to the present state of knowledge – the ion bombardment seems to play different roles for the different materials. The growth of c-BN relies, beyond doubt, also on attachment processes (which, however, take place below the surface – in contrast to diamond); the deposition of ta-C, on the other hand, relies on local densification or phase transition processes. Since the deposited material is crystalline in the first case but amorphous in the latter, it is not possible to compare both processes down to the last detail anyway.

The body of data existing at the present time is by far not sufficient to predict general mechanisms regarding the ion-assisted deposition of superhard materials in the B/C/N system, even if related materials such as B_4C or SiC[78] are also taken into consideration.

It is nevertheless possible to propose some general trends at this point. Ion-assisted deposition of superhard materials in the B/C/N system is evidently determined by two quantities, the ion bombardment B and the substrate temperature T_s. Both are distinguished by the fact that they bring about advantages as well as disadvantages from the point of view of the deposition of crystalline, high-density materials:

- The ion bombardment is evidently required in order to obtain the dense modification[79].
- On the other hand, ion bombardment always leads to the formation of defects, eventually even to amorphization.
- A high substrate temperature always supports the crystallinity of the material.

---

[78] In this case, in addition, the problem arises that SiC always exists in tetrahedral form; therefore, a densification step is not required.

[79] Nevertheless, the detailed mechanisms (selective sputtering, local densification, direct phase transition) are still open to discussion; they may even differ from material to material.

- On the other hand, high substrate temperatures could also cause relaxation processes favoring less dense materials[80].

The situation is further complicated by the fact that – with the exceptions of diamond and ta-C – binary or even ternary materials are always concerned, for which a stoichiometric composition evidently is a sine qua non for the formation of the dense modification. Since the substrate temperature and, especially, the ion bombardment have a strong influence on the stoichiometry[81], further constraints may arise rendering an optimization of T_s and B with respect to the crystallinity and modification difficult or, under certain circumstances, even impossible.

Table 3.8. Ionicity and minimum substrate temperature for the deposition of crystalline films under ion bombardment. This table does not by any means claim completeness but is, rather, intended to indicate some trends. For the ionicity, two different sets of values are given; the first refers to the definition (1.21) of Pauling, while the second relies on the papers of Phillips [80]; here, the data are taken from [26, 80]. Concerning the ionicity of SiC, there seem to be some uncertainties in the literature. According to [22, 31] it is significantly larger than that of β-C$_3$N$_4$, reaching the value of BN

Material	Diamond	B$_4$C	β-C$_3$N$_4$	SiC	c-BN	InN
Ionicity (Pauling)	0	0.06	0.08	0.13	0.25	0.46
Ionicity (Phillips)	0	≈ 0	0.15	> 0.18	0.38	0.58
$T_{\text{cryst,min}}$ (°C)	∞	900–1100	???	420	150	RT
Sources	—	[314]	—	[50]	[97, 26]	[26]

The above discussion can be illustrated by means of the data in Table 3.8. It shows for several materials the minimum substrate temperature required in order to deposit the dense modification in crystalline form under ion bombardment. The ionicities of the compounds are given also. Although the ion bombardments used for the various experiments summarized in Table 3.8 are by no means comparable, nevertheless a distinct trend becomes evident: the lower the ionicity, the higher the substrate temperature required for the deposition of crystalline films. If for any material at the appropriate temperature problems arise with respect to the maintainance of the densification or

---

[80] Even in the case of diamond, the formation of graphite takes place under deposition conditions at temperatures above 1000–1100°C, as has been discussed in Sect. 3.2.4. Bulk diamond graphitizes in the absence of air or in vacuum at about 1500°C [313].

[81] Section 3.5 shows that these effects are especially pronounced for β-C$_3$N$_4$. Nevertheless, preferential sputtering of nitrogen plays also a role in c-BN deposition [91, 99].

stoichiometry, ion-assisted deposition of this material resulting in crystalline layers seems to be impossible.

For example, on the basis of this table, it can be predicted that, owing to the low ionicity, relatively high substrate temperatures will be required for β-C_3N_4 in order to obtain crystalline material[82]. Since the effect of chemically enhanced preferential sputtering becomes more pronounced the higher the temperature, a successful deposition of crystalline β-C_3N_4 by means of ion-assisted techniques seems at the present time to be rather improbable.

---

[82] Indeed, Yu et al. [290] observed a slight increase of the (nevertheless in total very low) nanocrystalline fraction of their $C_{1-x}N_x$ films with increasing substrate temperature.

4. Nucleation Mechanisms

4.1 On the Theory of Nucleation

Nucleation denotes the formation of small clusters of film material on a substrate from individual atoms or molecules. The transition between the nucleation step and the following growth process is rather gradual; this is especially true in the case of layer growth discussed below, where the growth of each new layer requires a nucleation step. In order to describe the nucleation process, various formalisms have been developed [315, 316, 317, 318]: a macroscopic (phenomenological) thermodynamic description [315], a statistical thermodynamic description [315] and an atomistic description on the basis of rate equations [316, 317]. In the following, some of the most important aspects of these approaches will be presented briefly, as far as they are significant for the discussion of the nucleation of superhard diamond-like materials.

4.1.1 Macroscopic Thermodynamic Description

According to Kern et al. [315], the deposition of a film F on a substrate S can be regarded as the formation of a new phase. It depends decisively on the temperature T, the supersaturation p/p_0 (where p is the (partial) pressure of the condensing species and p_0 the equilibrium pressure of the forming film (crystal)) and the surface energies[1] $\sigma_i = (\partial G/\partial S_i)_{T,p,n,S_j,..}$, where S_j denotes the surface j.

On the attachment of a particle to the growing film, the chemical potential changes by

$$\Delta\mu = kT \ln \frac{p}{p_0} \; ; \tag{4.1}$$

if n particles are attached, the sum of the works done on the volume and surface can be expressed as

$$\Delta G(n) = -n\Delta\mu + \sum{}' \sigma_i S_i + (\sigma_{\text{int}} - \sigma_{\text{S}}) S_{\text{FS}} \; , \tag{4.2}$$

---

[1] Strictly speaking, the surface *free energies* are relevant here; in the following, the term *energy* is used for simplicity.

where the summation $\sum'$ has to be carried out over all free surfaces of the growing film (crystal); σ_S is the surface energy of the substrate and σ_{int} the interface energy

$$\sigma_{int} = \sigma_F + \sigma_S - \sigma_{ad} , \tag{4.3}$$

where σ_{ad} is the adhesion energy (i.e. the work gained if the surfaces F and S are brought into contact).

In thermodynamic equilibrium, all partial derivatives $(\partial G/\partial S_i)_{T,S_j,...,\Delta\mu}$ equal zero; this leads to the following conditions describing the equilibrium form of the crystal F on the substrate S:

$$\frac{\sigma_i}{h_i} = \frac{\sigma_j}{h_j} = ... = \frac{\sigma_{int} - \sigma_S}{h_{FS}} = \frac{\sigma_F - \sigma_{ad}}{h_{FS}} = \frac{\Delta\mu}{2v} . \tag{4.4}$$

Here v is the atomic volume and h_i the distance of the surface S_i from the center of the developing crystal; h_{FS} is the distance of the interface from the center; h_F denotes in the following the distance of the surface of the crystal (film) parallel to the interface from the center. Depending on the values of σ_{int} and σ_{ad}, the following cases have to be considered:

1. $\sigma_{ad} = 0 \longrightarrow h_{FS} = h_F$. There is no affinity between the film and substrate; the equilibrium form of the film corresponds to that of a crystal in the absence of a substrate.
2. $2\sigma_F > \sigma_{ad} > 0 \longrightarrow -h_F < h_{FS} < h_F$. A three-dimensional crystal is formed which is, however, truncated with respect to the equilibrium form without a substrate.
3. $\sigma_{ad} \approx 2\sigma_F \longrightarrow h_{FS} \approx 0$. In this case, only a two-dimensional film is formed.
4. $\sigma_{ad} \geq 2\sigma_F$. This case cannot be discussed on the basis of (4.2) and (4.4); rather, (4.2) has to be replaced by (4.7) (see below).

In the following, the two cases $\sigma_{ad} < 2\sigma_F$ (three-dimensional growth) and $\sigma_{ad} \geq 2\sigma_F$ (two-dimensional growth) will be discussed seperately.

Three-Dimensional Growth. Equation (4.2) can be rewritten in such a way that all terms contain explicitly the number of particles n of the crystal F:

$$\Delta G(n) = -n\Delta\mu + n^{2/3}X , \quad X = {\sum}' \sigma_i C_i + (\sigma_F - \sigma_{ad})C_{FS} . \tag{4.5}$$

Here, the C_i are geometrical constants ($S_i = n^{2/3}C_i$). If X is positive[2], $\Delta G(n)$ first increases with the number of particles n; however, in the case that $\Delta\mu > 0$ a maximum is reached (Fig. 4.1) where $\Delta G(n)$ starts to decrease again. This means that the crystal F has to reach a critical size n^*_{3D} before it starts to grow spontaneously with a decrease of $\Delta G(n)$. In this case, an activation barrier ΔG^*_{3D} has to be overcome:

---

[2] For $\sigma_{ad} < 2\sigma_F$, this is always the case.

$$n^*_{3D} = \left(\frac{2X}{3\Delta\mu}\right)^3, \quad \Delta G^*_{3D} = \frac{4X^3}{27\Delta\mu^2}. \tag{4.6}$$

From Fig. 4.1 and (4.5) it is evident that a supersaturation $\Delta\mu > 0$ is a sine qua non for this three-dimensional growth. The activation energy and critical cluster size are higher the lower the adhesion energy σ_{ad} is.

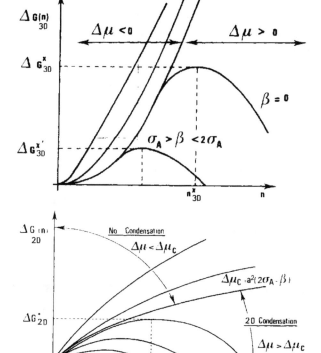

Fig. 4.1. Change of the free energy of a cluster as a function of the number of particles for the three-dimensional case: $\beta = \sigma_{ad} < 2\sigma_A = 2\sigma_F$ [315]

Fig. 4.2. Change of the free energy of a cluster as a function of the number of particles for the two-dimensional case: $\beta = \sigma_{ad} \geq 2\sigma_A = 2\sigma_F$ [315]

Two-Dimensional Growth. If $\sigma_{ad} \geq 2\sigma_F$, (4.2) has to be rewritten as follows[3]:

$$\Delta G(n)_{2D} = -n\Delta\mu + \sigma_F S_{FS} + (\sigma_F - \sigma_{ad}) S_{FS} - \sum_1^j \rho_i L_i. \tag{4.7}$$

The last term in (4.7) decribes the energies which are associated with the free ledges of the two-dimensional cluster (lengths of ledges L_i, free energies ρ_i). Again reformulating in such a way that all terms contain the number of particles n, one obtains (where the area of a particle is a^2)

---

[3] All free surfaces with the exceptions of S_F and the interface S_{FS} vanish.

$$\Delta G(n)_{2D} = -n\Delta\mu + n^{1/2} \sum_{1}^{j} \rho_i C_i + na^2(2\sigma_F - \sigma_{ad}) . \tag{4.8}$$

This case is depicted in Fig. 4.2. For $\Delta\mu > a^2(2\sigma_F - \sigma_{ad})$, i.e. $\Delta\mu > 0$, $\Delta G(n)_{2D}$ increases steadily, whereas a maximum appears for $\Delta\mu \leq 0$. This means that 2D growth, in contrast to 3D growth, can take place only at undersaturation. However, according to Kern et al. [315], the condition $\sigma_{ad} \geq 2\sigma_F$ is more important for 2D growth than this condition of undersaturation.

Summary. This macroscopic thermodynamic description thus leads to two different epitaxial modes of growth: a 3D condensation at supersaturation and a 2D condensation at undersaturation, which depend on the characteristic surface energies as follows:

$$\sigma_{ad} < 2\sigma_F \longrightarrow 3D , \qquad 2D \longleftarrow \sigma_{ad} > 2\sigma_F , \tag{4.9}$$
$$\sigma_F > \sigma_S - \sigma_{int} \longrightarrow 3D , \qquad 2D \longleftarrow \sigma_F < \sigma_S - \sigma_{int} . \tag{4.10}$$

4.1.2 Statistical Thermodynamic Description

The above considerations can be extended by statistical thermodynamic arguments [315]. Instead of the different surface energies, interactions between nearest neighbors are taken into account here:

$$\sigma_F = \frac{E_F}{2a^2}; \qquad \sigma_S = \frac{E_S}{2a^2}; \qquad \sigma_{ad} = \frac{E_{ad}}{a^2}. \tag{4.11}$$

In order to describe the interface, (4.3) is thus replaced by the relation

$$E_{int} = E_F + E_S - 2E_{ad} ; \tag{4.12}$$

the conditions for 2D and 3D growth, respectively, read accordingly

$$E_{ad} < E_F \longrightarrow 3D \qquad 2D \longleftarrow E_{ad} > E_F. \tag{4.13}$$

This means that three- or two-dimensional growth takes place if the interaction of a film atom with the substrate is weaker or stronger, respectively, than that with another film atom.

By means of statistical considerations of the occupation probability of neighboring nucleation sites, which for reasons of space will not be discussed here in detail (see [315]), and the resulting partition functions of the substrate, film and gas phase, so-called isotherms $\theta(p)$ and $\theta(\Delta\mu)$ are obtained, describing the mean coverage ratio θ of the substrate surface as a function of the pressure p and the chemical potential $\Delta\mu$, respectively (Fig. 4.3).

The statistical thermodynamic approach leads to the result that, besides the two- and three-dimensional growth processes already known, a third case exists; thus a total of three fundamental nucleation mechanisms has to be distinguished (Fig. 4.3):

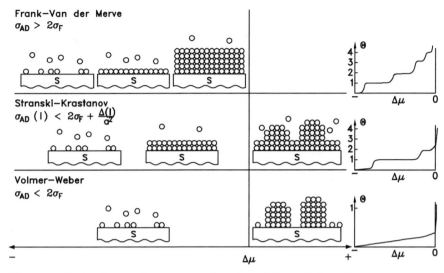

Fig. 4.3. Nucleation mechanisms as a function of $\Delta\mu$ and the adhesion energy $\sigma_{\rm ad}$ according to [315]. On the *far right*, the resulting isotherms are presented (θ = ratio of coverage)

- *Frank–van der Merwe growth* (layer growth, 2D growth): the interaction between the substrate atoms and the film (adhesion) is much stronger than the cohesion, i.e. the interaction of the film atoms with each other ((4.9) and (4.13)).
- *Volmer–Weber growth* (island growth, 3D growth): in this case, the interaction between the film atoms is stronger than the adhesion ((4.9) and (4.13)).
- *Stranski–Krastanov (SK) growth*. The SK growth represents a mixture of layer and island growth. It is observed, for example, if – after an original layer growth process – the adhesion energy decreases during the process because of changes of the surface with the number of deposited monolayers: $E'_{\rm ad} < E_{\rm ad}$. In this case, a transition to island growth takes place (Fig. 4.3). However, SK growth can also be caused by lattice distortions due to a mismatch between the film and substrate: in such a case, the films tries to adapt to the surface. To this end, for every monolayer j an elastic energy $\Delta(j)$ is required; the adhesion energy $E_{\rm ad}$ has therefore to be corrected by the elastic energy of the uppermost monolayer $\Delta(l)$ [315, 318].

4.1.3 Atomistic Description

In many cases, especially for the deposition of metals on insulators, the critical cluster size i is very low (only one or two atoms). A description of the nucleation process by means of thermodymanic quantities is therefore not appropriate; rather an atomistic (kinetic) approach is required [316].

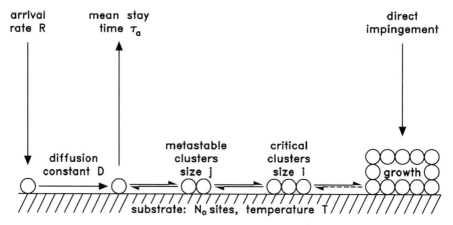

Fig. 4.4. Processes and parameters relevant to the nucleation of thin films according to [316]

The starting point is a substrate surface at a temperature T with N_0 adsorption sites of adsorption energy E_a [316, 318] (Fig. 4.4). Onto this surface, growth species (in the simplest case atoms of one species) impinge at a rate R which leads to a surface concentration n_1 of the atomic species. These atoms migrate on the surface with a diffusion constant D, which is determined by the diffusion energy E_d:

$$D = \frac{\alpha \nu}{N_0} \exp\left(-\frac{E_d}{kT}\right) . \tag{4.14}$$

(Here, ν is an atomic oscillation frequency ($\approx 10^{11}$–10^{12} Hz) and α a constant ($\approx 1/4$).) After a mean residence time τ_a, which is given by the adsorption energy such that

$$\tau_a^{-1} = \nu \cdot \exp(-\frac{E_a}{kT}) , \tag{4.15}$$

these species finally desorb again, unless they form beforehand – with other surface atoms or with already existing agglomerates – clusters of size j, which are characterized by a concentration n_j and binding energy E_j[4]. Hence, small clusters are unstable and may decay again; once a critical size i is reached, however, their further growth is more probable than their decay. These clusters, which may be either two-dimensional or three-dimensional (see below), continue to grow by capture of surface atoms as well as by direct attachment from the gas phase, until they coalesce and cover the complete surface.

---

[4] Reactions with or diffusion into the substrate are further possible loss mechanisms for surface atoms.

The nucleation process is therefore decribed by the parameters R, T_s and N_0 as well as by the adsorption energy E_a, the diffusion energy E_d and the binding energies E_j of clusters of size j. Depending on the ratios of these quantities to each other, the three growth modes discussed above again result [316][5]; however, the transitions between them are rather gradual.

The nucleation process itself, finally, can be decribed by the following rate equations:

$$\frac{dn_1}{dt} = R - \frac{n_1}{\tau_a} - 2U_1 - \sum_{j=2}^{\infty} U_j \quad \text{and} \quad \frac{dn_j}{dt} = U_{j-1} - U_j . \quad (4.16)$$

Here, the terms on the right-hand side of the first equation represent the arrival rate of atoms, their desorption, the formation of two-atom clusters and the (net) attachment rate of an atom to a cluster of size j, respectively. The second equation takes into account the fact that the density of clusters of size j can increase by attachment of an atom to a cluster of size $j-1$, and decrease by attachment to a cluster of size j.

For reasons of space, we refrain from a discussion of the solutions of these rate equations for the different cases and only refer to the appropriate literature [316, 317].

4.1.4 Influence of Surface Defects

Up to now, only ideal surfaces have been considered. Defects, on the other hand, may play a most important role in nucleation processes. In many cases in which on ideal surfaces only a rather low nucleation density is found, high concentrations of nuclei are observed at defects such as steps, etching grooves, dislocation spirals and point defects ('decoration' of defects). Defects may, among others things [316],

(1) represent nucleations sites with adsorption energies $E_{ad}^d > E_{ad}$,
(2) enhance the binding energies of clusters or
(3) influence the diffusion constants.

An enhancement of the nucleation density due to defects is observed especially in those cases in which the adsorption energy is low with respect to the binding energy E_F [316]. For diamond-like materials this condition is fulfilled for almost all substrates, especially for the case of silicon; as a consequence, defects play a most important role especially for the nucleation of diamond as will be discussed in Sects. 4.2.2 and 4.2.4.

---

[5] A classification presented by Venables and Price [316] distinguishes a number of subcases for each mode. For example, for $R/N_0 D \geq 1$, only amorphous growth is possible since diffusion processes can hardly take place; for $E_2 \ll kT$, on the other hand, nucleation on clean surfaces is almost impossible (see below).

4.1.5 (Hetero)epitaxy

The case of heteroepitaxy will be addressed briefly, i.e. the growth of single-crystal films with a fixed crystallographic relationship to the substrate. Although heteroepitaxy is in principle possible for each of the nucleation modes, when considering the underlying mechanisms one has to distinguish between layer and island growth. In the case of layer growth, epitaxy takes place almost always (if crystallographically possible), as long as the substrate is clean (perfect) enough. The influence of the substrate is so strong that 'the growing film has little option but to choose the best (i.e. necessarily epitaxial) orientation in which to grow' [316]. For island growth, epitaxy is also possible; nevertheless, at least in part this has to be regarded as a postnucleation phenomenon. On the one hand, the enthalpy of formation is in general lower for oriented nuclei than for misoriented ones [318]. Owing to the condition $E_{ad} < E_F$, on the other hand, even stable clusters (nuclei) remain movable; thus they can reorient themselves even after their formation or may even be able to migrate [316]. Finally, recrystallization effects may occur during the coalescence of the individual clusters.

Whether a film grows as single-crystal, polycrystalline (eventually textured) or even only amorphous thus depends on the materials involved, on the substrate temperature and on the growth rate: for every deposition rate R a critical temperature T_e exists above which epitaxial growth takes place. Namely [318],

$$R < A \exp\left(-\frac{Q}{kT_e}\right). \tag{4.17}$$

Finally, the influence of the mismatch $\Delta = (a_s - a_f)/a_f$ has to be discussed briefly; this is present in most cases of heteroepitaxy and is caused by different lattice parameters a_s and a_f of the substrate and film, respectively [318]. For small Δ and thin layers, the film adapts to the surface by the development of strain (pseudomorphic growth). If a certain thickness is exceeded, the introduction of misfit dislocations together with a simultaneous reduction of the strain is energetically more favorable. For very thick layers, the misfit is compensated by misfit dislocations as well as by strain; thus, with increasing thickness the lattice parameter of the bulk material is gradually reached by the increasing incorporation of dislocations.

For large misfits, however, so-called coincidence lattices are formed where not only the film lattice but also that of the substrate relaxes at the interface. The mismatch of the coincidence lattice $(ma_s - na_f)/a_f$ is then again compensated by strain and misfit dislocations.

In general, however, the extent of the misfit between the substrate and film is, according to Reichelt [318], not decisive for the feasibility of heteroepitaxial growth; rather, the different energies involved, the temperature and the growth rates play more important roles, as we have seen above.

4.1.6 Nucleation of Superhard Materials

In order to fully exploit all of the extreme properties of diamond-like superhard materials described in Chap. 2, the deposition of single-crystal (and moreover almost defect-free) films is required, i.e., in other words, heteroepitaxial growth. This is especially true for electronic and optoelectronic applications. Such heteroepitaxial growth can be most easily achieved by layer growth; however, as discussed above, it is also possible in the case of island growth if all nuclei have a fixed crystallographic relationship among each other and with the substrate, thus allowing the coalescence of nuclei without the formation of grain boundaries[6]. Owing to the high surface energies of diamond and c-BN[7], layer growth seems impossible to reach; in order to achieve heteroepitaxy, island growth with the formation of oriented nuclei is therefore required.

For a number of applications (e.g. for tribological purposes), however, the deposition of polycrystalline films as already achieved at the present time by low-pressure diamond synthesis, is sufficient. Nevertheless, in the case of island growth a high nucleation density is required for single-crystal as well as for polycrystalline growth. The lower n_{nuc} is, the higher is the minimum thickness of continuous films because of the three-dimensional growth of isolated crystals. Also, the surface roughness, which is of major importance for optical applications, for example, is decisively influenced by the nucleation density. The film adhesion, finally, is in general the better the higher n_{nuc} is.

The optimum conditions for the growth of a thin film are not necessarily the best for its nucleation nor vice versa. For diamond this is indeed the case, as will become evident in the following sections, especially during the discussion of bias-enhanced nucleaction (BEN) (Sect. 4.3), whereas in the case of c-BN this question is still open ([204]); nevertheless, the results presented in Sect. 3.3.5 also hint in this direction.

In the following, the nucleation mechanisms of the four materials discussed in this book are examined, thereby restricting the discussion to a large extent to silicon as the substrate material; other substrates are only taken into account in those cases in which the respective experiments provide new and important information. It will become evident that classical nucleation, as discussed in the preceding section (i.e. the formation of critical nuclei on a foreign but well-defined substrate), plays a role only in some special cases of diamond nucleation. In general, the deposition of the superhard diamond-like materials considered in this book starts with a distinct modification of the substrate surface due to the special deposition conditions, before – nevertheless by mechanisms which are different for each material – the first crystallites or fractions of the desired modification are formed.

---

[6] In this context, compare the epitaxial growth of diamond on c-BN (Sect. 4.2.1) and the BEN (Sect. 4.3).

[7] In [8] the following values are, for example, given: diamond $\langle 111 \rangle$, 5.4 J/m^2; silicon $\langle 111 \rangle$, 1.2 J/m^2; graphite $\langle 0001 \rangle$, 0.07 J/m^2.

4.2 Nucleation of Diamond

4.2.1 General Considerations

The nucleation of diamond during low-pressure synthesis is, by far, not as well understood as its growth [319, 320]. It it generally accepted at the present stage of knowledge that it depends on a combination of suppression of graphite nuclei, selective etching of sp^2 carbon and stabilization of diamond nuclei and surfaces [76]. It is, moreover, evident that, as in the case of growth, atomic hydrogen plays a decisive role. This does not only relate to selective etching and the stabilization of surfaces; Badziag et al. [123] show that hydrogen-terminated diamond nanocrystals are more stable than the corresponding graphite crystallites, in contrast to the bulk materials. It is, therefore, not possible to describe the nucleation of diamond simply in terms of the atomistic model presented above; for thermodynamic considerations also, the role of atomic hydrogen has always to be taken into account.

The situation is further complicated by the observation that the nucleation of diamond always proceeds as a two-step process, for the details of which one has nevertheless to distinguish between carbide-forming (Si, Mo) and non-carbide-forming substrates (Ni, Pt): for carbide-forming materials, diffusion of carbon into the substrate and subsequent carburization always takes place, until finally the carbide layer thus formed acts as a barrier to further carbon diffusion. This barrier then leads to an increase of the carbon concentration on the surface (supersaturation). This barrier, however, is only a necessary, not a sufficient, condition for rapid formation of nuclei. For non-carbide-forming substrates, on the other hand, deposition starts with the formation of a graphitic, or more, generally, a hydrocarbon layer before the first diamond nuclei can be observed [171, 321, 322]. In general, the nucleation rate on non-carbide-forming substrates is much lower than on materials with a tendency to build up carbide layers.

From the above it is evident that in every case, prior to the formation of diamond nuclei, a substantial modification of the substrate surface takes place (formation of a carbide or deposition of a hydrocarbon layer). It is therefore generally agreed that two conditions have to be fulfilled for the nucleation of diamond [76]:

1. carbon supersaturation of the surface
2. existence of either nucleation sites or seed crystals.

The specific nature of growth sites suited for diamond nucleation will be discussed in the following sections.

Finally, the special case of heteroepitaxy has to be addressed. In the literature there exist a limited number of studies concerning the epitaxial growth of diamond layers on foreign substrates [327]. Hereb, the most important aspects for the selection of the substrate material are a lattice-matching to

Table 4.1. Data relevant to heteroepitaxy of diamond and c-BN on various substrate materials. References are given only for those exotic substrate/film combinations which are not further covered in the text. Copper has been proposed for heteroepitaxy because of its lattice parameters; owing to its very low surface energy, however, this aim could not be achieved. ZB, = zinc blende structure; WZ, wurtzite structure. The mismatch Δ was calculated according to $\Delta = (a_s - a_f)/a_f$ or $\Delta = (na_s - ma_f)/ma_f$ in the case of an $n : m$ matching. For the cubic/hexagonal combination of systems, the first value concerns the mismatch with respect to the c axis, the second that normal to it. The indices 's' and 'f' refer to the substrate and film, respectively

Material	Crystal structure		Δ	Crystallographic relations		References
	a (nm)	c (nm)	[%]			
Heteroepitaxy of diamond films						
Diamond	ZB 0.3567		—	—		
c-BN	ZB 0.3616		1.4	$(100)_s \parallel (100)_f$	$[001]_s \parallel [001]_f$	[323]
			1.4	$(111)_s \parallel (111)_f$	$[\bar{1}10]_s \parallel [\bar{1}10]_f$	[324]
β-SiC*	ZB 0.4358		22	$(100)_s \parallel (100)_f$	$[011]_s \parallel [011]_f$	
Silicon*	ZB 0.5430		52	$(100)_s \parallel (100)_f$	$[011]_s \parallel [011]_f$	
Graphite	hex. 0.246	0.671	8.6/−2.5	$(0001)_s \parallel (1\bar{1}1)_f$	$[11\bar{2}0]_s \parallel [110]_f$	
α-SiC	WZ 0.324	0.521		$(0001)_s \parallel (1\bar{1}1)_f$	$[11\bar{2}0]_s \parallel [110]_f$	[325]
BeO	WZ	0.438	6.2/—	$(0001)_s \parallel (1\bar{1}1)_f$	$[11\bar{2}0]_s \parallel [110]_f$	[324]
Nickel	f.c.c. 0.352		−1.3	$(100)_s \parallel (100)_f$	$[001]_s \parallel [001]_f$	[326]
Copper	f.c.c. 0.361		1.2	—	—	—
Heteroepitaxy of c-BN films						
c-BN	ZB 0.3616		—	—		
h-BN*	hex. 0.2504	0.330		$(0001)_s \parallel (1\bar{1}1)_f$	$[11\bar{2}0]_s \parallel [110]_f$	

* Heteroepitaxial growth is only possible with BEN or ion-assisted processes.

diamond as good as possible and a surface energy as high as possible; additionally, the thermal expansion coefficient should not differ too much from that of diamond [324, 328]. The materials investigated up to now are compiled in Table 4.1.

With regard to all these points, c-BN is the most promising candidate for diamond heteroepitaxy. Indeed, epitaxial growth of diamond can be rather easily achieved on {111} as well as on {100} c-BN surfaces [323, 324, 329][8]. However, since at the present time large-area, single-crystal c-BN substrates

---

[8] This is, however, only true for boron-terminated surfaces; it is not true for nitrogen-terminated surfaces; compare Sect. 2.2.2.

are not available, heteroepitaxy of diamond on c-BN is indeed of great importance from a scientific point of view but technologically rather without relevance.

Epitaxial growth of diamond has also been observed on some other materials such as α-SiC, BeO and Ni; nevertheless, most of these substrates are likewise of minor technological importance. This is not the case for nickel, which in addition has only a small lattice mismatch to diamond; nickel, however, supports catalytically the formation of graphite [327]; as a consequence, diamond growth on Ni has turned out to be very difficult to achieve, although in some cases heteroepitaxial growth has been accomplished [326].

At the present time, therefore, only Si and β-SiC present themselves as substrates which are, on the one hand, of technological relevance, relatively cheap and available in large numbers, and on which, on the other hand, diamond heteroepitaxy seems to be very promising even if it has not been finally established yet; heteroepitaxial growth on these substrates, however, can only be achieved by means of bias-enhanced nucleation (BEN), which will be discussed separately in Sect. 4.3.

4.2.2 Nucleation on Untreated Substrates

In general, the nucleation density of diamond on foreign substrates is very low; on polished silicon substrates, for example, it is only 10^4 cm^{-2}.[9] In most cases the formation of the first nuclei is additionally preceded by a long incubation time, which – depending on the substrate – can even last several hours [76]. Finally, on untreated substrates the nucleation takes place almost exclusively at surface defects (Fig. 4.5). All these observations show that the formation of the first diamond nuclei is very difficult to achieve; this means that the nucleation proceeds by no means as spontaneously as the subsequent growth.

In agreement with these observations is the fact that the process of diamond nucleation on foreign substrates is not easily comprehensible from a theoretical point of view [55, 320], since, on the one hand, diamond is metastable with respect to graphite and, on the other hand, all principle surfaces possess extremely high surface energies. According to (4.9), layer growth seems therefore to be impossible on any foreign substrate; in addition, according to (4.6) the critical size of stable diamond nuclei is very high.

However, again it has to be taken into account that the phase diagram in Fig. 2.1 concerns the carbon system, not the relevant (atomic) hydrogen/carbon system. Likewise, the surface energies given in the literature are related to clean, not to hydrogen-terminated diamond surfaces.

Accordingly, most of the attempts to explain the nucleation of diamond are based on the presence of high concentrations of atomic hydrogen at the

---

[9] By means of the pretreatment methods described in the following, on the other hand, nucleation densities of about 10^9–10^{11} cm^{-2} can be achieved.

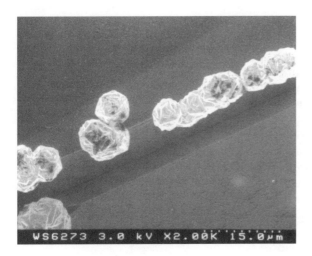

Fig. 4.5. Nucleation on an anisotropically etched V-shaped trench on a silicon surface. Diamond crystallites are observed neither on the plane substrate nor on the concave edges, but only at the convex protrusions

surface [54, 55, 76, 320]. In Sect. 3.2.2 the study of Badziag et al. [123] has already been mentioned, which shows that hydrogen-terminated diamond nanocrystals ($<$ 3 nm) are more stable than graphite crystallites of the same size, in contrast to the corresponding bulk materials.

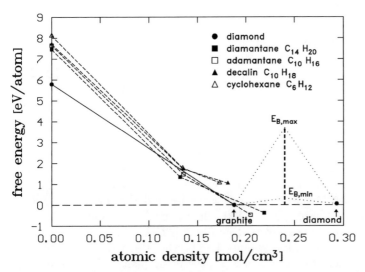

Fig. 4.6. Free energies of gas phase atoms, some aromatic unsaturated sp^2 compounds and the corresponding saturated sp^3 compounds according to Angus et al. [320]. In addition, data for graphite, diamond and (estimated) extreme values of the energy barrier between the two modifications are indicated. The reference point is the system graphite + H_2 at 1000 K

These considerations have been extended by Angus et al. [320]. In Fig. 4.6, which is based on the data given in [320], values of the free energy for some systems for free gas phase atoms, unsaturated aromatic sp^2 species and saturated sp^3 species are presented[10]. Additionally, the values for graphite, diamond and the extreme values of the energy barrier between them (as given by Angus et al.) are also presented in the figure.

From Fig. 4.6 it is again evident that in the hydrogen-free case, graphite is energetically more favorable than diamond; thus there is no reason for a transformation from graphite to diamond. On the other hand, the figure shows also that in the presence of atomic hydrogen the free energies favor a subsequent reaction: gas phase species $\longrightarrow$ aromatic sp^2 species $\longrightarrow$ saturated sp^2 species (thereby, diamantane and adamantane are energetically more favorable than graphite). Presumably there also exist energy barriers, which, however, according to Angus at al., should not be very high in the presence of atomic hydrogen.

Thus, according to Angus et al., the nucleation of diamond takes place via graphitic or aromatic precursors, which are eventually formed on the substrate surface because of the carbon supersaturation and which – because the process is thermodynamically favored – are hydrogenated by atomic hydrogen, thereby forming diamond precursors. Such graphitic or aromatic structures have indeed been observed experimentally during the nucleation of diamond [321, 322].

4.2.3 Nucleation on Highly-Oriented Graphite

The nucleation of diamond on highly oriented pyrolytic graphite (HOPG) represents an interesting but special case. Although it is without relevance from a technological point of view, it may eventually contribute to an understanding of the nucleation mechanisms not only of diamond but also of c-BN.

Li et al. [330] observed that diamond nucleates preferentially on the prism planes of HOPG but not on the basal planes. In addition they found a preferential orientation of the diamond nuclei, with the diamond (111) planes parallel to the (0001) basal planes of graphite and the diamond $[1\bar{1}0]$ direction parallel to the graphite $[11\bar{2}0]$ direction. These observations have, meanwhile, been confirmed several times (e.g. [325]).

On the basis of these observations, Lambrecht et al. [319] have proposed a model for the nucleation of diamond on the prism planes of graphite which takes into account also the following facts:

---

[10] For example, the diamantane system consists of the the following: gas phase, 14C and 20H; sp^2 species, phenanthrene $C_{14}H_{10}$ + 10H; diamantane, $C_{14}H_{20}$. It should be mentioned that adamantane and diamantane are the smallest completely hydrogenated molecules with the structure of diamond.

1. In the graphite (0001) and in the diamond (111) plane the atoms are arranged in hexagons; however, the diamond (111) plane is not flat but, rather, 'buckled'. Nevertheless, in projection both planes show approximately the same geometrical arrangement (see Fig. 4.20).
2. The spacings of the graphite (0001) and the diamond (111) planes are in the ratio 2:3 (see Table 4.1).
3. The densities of graphite and diamond are approximately in the ratio 2:3 (2.26 g cm^{-3} : 3.52 g cm^{-3}).

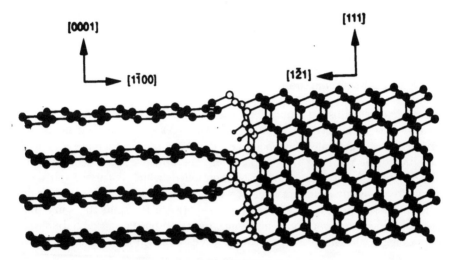

Fig. 4.7. Nucleation structure of diamond on graphite according to [319]

On the basis of these facts and observations, Lambrecht et al. postulated a 2:3 lattice-matching as the basis of the nucleation of diamond on the prism planes of graphite, as depicted in Fig. 4.7. According to their molecular-dynamic calculations, this structure possesses a very low interface energy σ_{int} of only ≈ 1.7 Jm^{-2}. The most important contribution to σ_{int} originates from the non-saturated bonds (dangling bonds) of every third diamond layer. Saturation of these bonds by hydrogen further lowers the interface energy (≈ 0.4 Jm^{-2}). The hexagonal network continues from the graphite into the diamond without a noticeable contribution of strains to the total interface energy.

Under the special conditions of diamond deposition, the nucleation will be initiated by hydrogen saturation of the dangling bonds of the prism planes of graphite[11]. This hydrogen saturation leads to a bending of the basal planes at their ends; by hydrogen abstraction and subsequent incorporation of hy-

---

[11] In contrast, all bonds within the basal planes of graphite are saturated; as a consequence, these planes are not attacked by atomic hydrogen

drocarbon species (in analogy to the growth of diamond), the basis is laid for the third plane which is required for the growth of diamond[12].

Recent molecular-dynamic calculations of Jungnickel et al. [331] confirm to a large extent the conclusions of Lambrecht et al. Although slight corrections with respect to the interface energy given above and some details of the structure shown in Fig. 4.7 seem to be necessary, the adhesion energy

$$\sigma_{ad} = \sigma_D + \sigma_G - \sigma_{int}, \tag{4.18}$$

(where $\sigma_{D,G}$ represent the surface energies of the relevant diamond and graphite surfaces, respectively, and σ_{int} the interface energy (cf. Sect. 4.1.1) is still very large; therefore, the structure proposed by Lambrecht et al. seems to be stable even at the elevated temperatures used during the deposition of diamond.

4.2.4 Abrasive Treatment

Since on most substrate materials the diamond nucleation density is extremely low and since, in addition, nucleation requires a long incubation time, various techniques have been developed in order to increase n_{nuc} while simultaneously reducing t_{nuc}. First of all, the abrasive treatment with diamond powder (scratching) has to be mentioned, which is carried out either manually or by ultrasonic treatment with a liquid suspension of the powder. With both techniques, nucleation densities between 10^{10} and 10^{11} cm^{-2} have been obtained on a variety of materials; owing to better reproducibility and homogeneity, the ultrasonic treatment has established itself as the standard technique.

The discussion of the nucleation mechanisms relevant to the abrasive treatment has not yet been completed [76]. Nevertheless, it can be taken for certain that the creation of defects on the substrate surface (compare Sect. 4.2.2) by the diamond crystallites plays an important role [332, 333][13]; of even greater relevance, however, is the fact that after the pretreatment diamond particles with diameters between 5 and 50 nm remain in or on the surface [322, 333, 334]. These have a density between 10^{10} and 10^{11} cm^{-2} which corresponds to the later nucleation density, and are of irregular shape and random orientation. These particles then serve as nuclei for the subsequent growth of diamond [333, 334]. This means that during the abrasive treatment no nucleation in the true sense of this term but, rather, homoepitaxial growth on the particles left behind by the treatment takes place.

---

[12] Compare the interface region in Fig. 4.7, which is emphasized by the open symbols.

[13] This is confirmed by the fact that not only diamond but also c-BN and SiC powders can be used for the treatment; nevertheless, here the incubation time is longer [76] and the treatment in general not as effective [333] as in the case of diamond powder.

Thus it has to be concluded that the effect of the abrasive treatment – irrespective of the weights of the two processes discussed in the preceding section – can be traced back to mechanisms (nucleation at defects/homoepitaxy on residual diamond particles) already well-known. It is also evident that this technique, while indeed increasing the nucleation density drastically, nevertheless always leads to randomly oriented nuclei, thus rendering heteroepitaxial growth impossible.

4.3 Bias-Enhanced Nucleation of Diamond

4.3.1 General Considerations

High nucleation densities similar to those in the case of the abrasive treatment discussed above can be obtained by means of the bias-enhanced nucleation (BEN) of diamond. In 1991 Yugo et al. [335] observed that during MWCVD the nucleation density on silicon substrates can be increased drastically by applying a negative bias voltage between the plasma and substrate in the starting phase of the deposition[14]; by means of the BEN process, nucleation densities on the order of 10^9–10^{10} cm^{-2} are possible. Typically, the applied voltages are in the order of -200 V; the other parameters are mainly identical to those for the standard deposition conditions, with the only exception that in most cases the methane fraction in the gas phase is slightly increased (2–10 % instead of 0.5–2 %). Compared to the abrasive treatment, BEN bears the advantage that it can be carried out in the same reactor as the subsequent deposition; additionally, the cleaning steps necessary after abrasive treatment are avoided. On the other hand, up to now BEN has been almost exclusively limited to MWCVD, and reports on its application during HFCVD are still scarce [336, 337, 338]; nevertheless, it seems to be possible in principle. Si and SiC are typically used as substrates; increased nucleation densities as a consequence of a BEN treatment have nevertheless been observed also on a variety of carbide-forming refractory metals [339]; heteroepitaxial growth has also been observed on nitrogen-terminated {111} c-BN surfaces [340][15] and iridium [341].

However, when compared to the abrasive pretreatment, BEN has another, most important adavantage: while the former inevitably leads to randomly oriented diamond nuclei (see above), the BEN process results, under certain conditions, in diamond nuclei which are oriented with respect to each other

---

[14] If the bias voltage is applied during the entire deposition process, only layers with a high fraction of non-diamond-bonded carbon are obtained (compare in this context the discussion of the influence of ion bombardment on the deposition of diamond in Sect. 3.3.6).

[15] In contrast to boron-terminated {111} surfaces, the nucleation density on nitrogen-terminated {111} c-BN surfaces is low without the application of BEN, and heteroepitaxy is not possible [323].

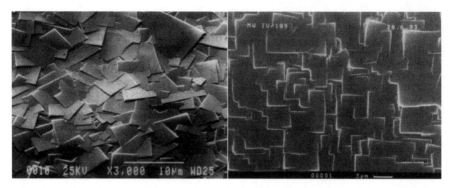

Fig. 4.8. SEM images of two diamond layers which have been deposited with a growth parameter α of about 2.95. This results in (100) planes which are almost parallel to the substrate surface but nevertheless in most cases possess a tilt angle of a few degrees. The sample on the *left-hand side* has been subjected to abrasive treatment; apart from the fiber texture, the individual crystals are randomly oriented. The sample on the *right-hand side* has been pretreated by BEN; the crystals possess an orientation with respect to each other which nevertheless is still imperfect. From [178]

and also to the underlaying substrate (see Fig. 4.8). This has been demonstrated independently for silicon substrates by Jiang et al. [342] and for β-SiC substrates by Stoner et al. [343]. In combination with the textured growth of diamond films discussed above, this oriented nucleation is an important step towards heteroepixial growth of diamond layers[16].

Despite the enormous advantages of BEN compared to the abrasive treatment and despite the resulting intensive research activities in this field, the BEN process still has a number of serious problems which have to be solved before an industrial application of BEN may be realized [344, 345]:

1. Very often, the nucleation density is inhomogeneous; in addition, in most cases the area on which a high nucleation density is obtained is considerably smaller than the area on which, in a given reactor, homogeneous diamond layers can be deposited [346, 332, 347, 348, 349, 350, 351].
2. The reproducibility of BEN processes is poor at present. This is true for a given reactor from process to process, and also for the transfer of a given process from reactor to reactor [344, 352].

These two aspects are of concern for all applications of BEN; improvement of homogeneity and reproducibility is a sine qua non for every application of

---

[16] Films such as that shown in Fig. 4.8 are often erroneously labeled epitaxial. This term, however, refers exclusively to single-crystal layers; the diamond layers obtained up to now, however, are still polycrystalline. Nevertheless it seems to be justified to speak of (almost) epitaxial growth of individual crystallites. Thus the term 'heteroepitaxial growth' may be applied but not the expression 'heteroepitaxial layers'.

the BEN process on an industrial scale. However, with respect to oriented nucleation, i.e. from the point of view of possible heteroepitaxial deposition, another most important problem has to be addressed:

3. The orientation of the nuclei and, consequently, of the resulting crystallites with respect to each other and to the underlaying substrate is still imperfect; rather, tilt and rotational angles are observed which are randomly distributed [328, 353, 354, 355] (compare Fig. 4.8). Schreck and Stritzker have shown recently [328] that the rotational deviations are a consequence of the BEN process itself; by improvements of their process, Jiang et al. were able to reduce the full width at half maximum (FWHM) of the distribution of the rotational deviations from 12° [342] to 4.6° [345] and, finally, to less than 2° [356][17].

Solutions of the problems addressed above require a detailed understanding of the elementary mechanisms relevant to BEN. Despite the intensive investigations carried out up to now, however, this aim is far from having been achieved. The problem can be stated as follows:

4. At the present time no model exists describing the mechanisms of BEN correctly even in a first approximation. Although a number of attempts have been made to explain the enhanced nucleation density as well as the oriented growth of nuclei, these are, however, at best qualitative and very often just speculative. Among the effects proposed to be responsible are changes of gas phase chemistry [353, 358], changes of surface chemistry (e.g. enhanced surface mobilities, enhanced surface reactivities and increased sticking probabilities of carbon species) [342, 355], creation of nucleation sites [350], creation of defects [345] and, finally, ion-induced mechanisms [359, 360, 361].

Robertson [360] and Gerber et al. [361, 362] have attempted to apply the subplantation model, which has been discussed above in the context of the deposition of c-BN and ta-C, to BEN; as will become evident from the discussion below, however, this model in its present formulations is not suited to describe the existing experimental observations correctly.

A complete characterization of the BEN process, which is a necessary condition in order to obtain an in-depth understanding of the underlying mechanisms is, however, rendered difficult by experimental problems:

5. Owing to the special conditions of diamond deposition, the internal process parameters depend in a complex manner on the external parameters, which in addition are, at least in part, interdependent. A complete control of the internal parameters is therefore, as will be discussed below,

---

[17] Very recently, Schreck et al. [357] were able to reduce the rotational misalignment considerably for $SrTiO_3$/Ir substrates, thereby reaching almost heteroepitaxial layers. If these results can be confirmed and improved, they may lead to a major breakthrough.

not possible at present. As a consequence, the parameters relevant to BEN have not yet been identified as has been, for example, the case with the growth of c-BN.
6. According to present knowlegde, diamond nuclei develop from (hydro)carbon particles with dimensions in the nanometer range. Therefore, characterization also on a nanometer scale is required. This is, however, beyond the scope of standard techniques for the characterization of diamond such as Raman, SEM, XRD and FTIR. The modern scanning-probe microscopies such as atomic force microscopy (AFM) and scanning tunneling microscopy (STM) would be most suited; owing to the special properties of diamond (e.g. hardness, conductivity), however, their application is by no means straightforward [363]. Thus, at present the very time-expensive transmission electron microscopy (TEM) is the most promising technique.

Owing to these difficulties, the present knowledge concerning the mechanisms of BEN mainly originates from indirect conclusions rather than from direct observations.

In the following, first the existing facts will be presented, making especial use of the investigations of BEN performed by our group. Subsequently, possible mechanisms and the models proposed up to now in the literature are discussed.

4.3.2 Experimental Observations

Electrical Characterization. Figure 4.9 shows the dependence of the current flowing between the plasma and substrate holder during a BEN process on the bias voltage applied.

To begin with, from the figure it is evident that the current depends on the nature of the substrate surface; in the entire voltage range it is higher for a diamond-covered substrate than for a clean silicon substrate. Both I/V curves can be roughly divided into three regions [364]: a region of mobility-limited current (I), a saturation region where every ion entering the plasma sheath will reach the substrate (II) and, finally, a strong increase of the current (III), which is caused by additional ionization processes in the sheath due to the applied voltage [364]. Typically, BEN is carried out in this third region.

The enhancement of the current in the case of diamond-covered surfaces is due to the enhanced emission of secondary electrons from diamond surfaces [364, 352, 365][18].

Since the current between the plasma and substrate depends on the nature of the substrate surface, it will necessarily vary with time during a BEN process, which starts with a clean silicon substrate and ends up with the

---

[18] Similarly, Hahn et al. observed a current increase during DC sputter deposition of c-BN after nucleation of c-BN, which has been attributed to the higher secondary-electron yield of c-BN compared to h-BN and Si [366].

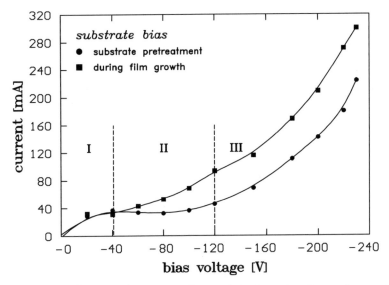

Fig. 4.9. I/V curve for a pure silicon substrate and a diamond covered substrate during a BEN process

formation of diamond nuclei. Indeed, a characteristic time dependence $I(t)$ is observed, which is shown in Fig. 4.10 and which will be addressed in detail in the discussion below.

The reproducibility problems mentioned above are, in addition, closely connected to the high secondary-electron coefficient of diamond, as we have shown in a recent investigation [344], which for reasons of space will not be discussed here in detail: diamond-covered surfaces on the substrate holder and on the shields in the vicinity of the holder have a considerable influence on the plasma and its geometrical shape; frequently, secondary plasmas are observed in the vicinity of such surfaces. Since the area covered with diamond and also the quality of the diamond film may not be constant, differences can occur from process to process.

In a similar manner, the problem of the limited coating area can be explained [344]. The conditions during a BEN process correspond to those of a normal glow discharge [367]; this means that at constant voltage a change of the discharge current can take place by a variation of the effective electrode area [220]. On application of the bias voltage the microwave plasma ball contracts, while simultaneously reaching down to the substrate. As soon as diamond nucleates in the middle of the substrate, the plasma ball enlarges steadily, and the nucleation front migrates from the middle of the substrate to its rim. Typically, however, the BEN step is terminated earlier since otherwise the quality of the already created nuclei degrade. This migration of the nucleation front is, however, again strongly influenced by diamond-covered surfaces in the vicinity of the substrate (compare in this context also [349]).

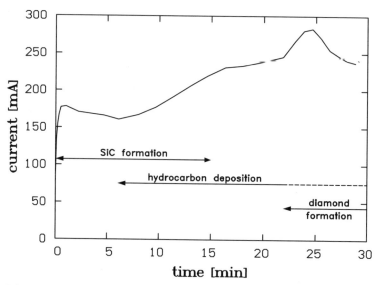

Fig. 4.10. Dependence of the current during a BEN process on time. The figure also includes the individual phases of the process, evaluated from the FTIR analysis

Gas Phase Investigations. In the initial phase of investigation of the BEN process, speculations on possible mechanisms concentrated especially on chemical processes (in analogy to the growth of diamond). Enhancement of the surface mobility and the sticking probability of growth species, or of the reactivity of these species by the applied voltage have been discussed [342, 355], as well as changes of gas phase chemistry. Shigesato et al. [358], for example, proposed that the concentration of atomic hydrogen in the vicinity of the substrate is increased by the application of the bias voltage; this atomic hydrogen then serves to stabilize diamond nuclei.

In order to test this assumption we have carried out plasma diagnostic measurements by means of optical emission spectroscopy (OES) [106, 364]. The most important results of these measurements are presented in Fig. 4.11, showing the dependence of the relative concentration of atomic hydrogen [H•] in the direct vicinity of the substrate on the applied bias voltage. At the transition to region III of the I/V curve (Fig. 4.9), [H•] decreases by approximately 25%. This is true for clean silicon substrates as well as for diamond-covered surfaces; nevertheless, the transition to region III depends on the nature of the substrate surface, as is already evident in Fig. 4.9.

Similar measurements have also been carried out by Sheldon et al. [368]. At 50 mbar, they observed an increase of [H•] by approximately 25% when the bias voltage was applied, but almost no changes at 35 mbar. Our own measurements were performed at 26 mbar (25% decrease). It can therefore be concluded that indeed a slight influence of the bias voltage on the concen-

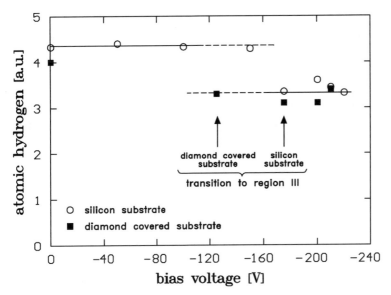

Fig. 4.11. Dependence of the concentration of atomic hydrogen on the applied bias voltage. The dependence of the transition to region III (Fig. 4.9) on the substrate material is reflected again by these OES measurements

tration of atomic hydrogen exists, which in addition depends on pressure[19]; nevertheless this slight influence seems not to be sufficient to explain the drastic increase of the nucleation density due to BEN.

Thus our OES measurements show, in summary [106, 364], that the gas phase composition in the vicinity of the substrate is not changed drastically by the application of the bias voltage. Also, the electron temperature remains approximately constant; only the electron density increases in region III by a factor of two, in agreement with Fig. 4.9.

Parameter Dependences. Two experiments decribed recently in the literature, by McGinnis et al. [348, 370], and by Schreck and Stritzker [328] and Jiang et al. [371], prove beyond doubt that BEN relies on direct ion bombardment of the growing film (Fig. 4.12).

On structured substrates, an increase of the nucleation density is observed only on surfaces parallel to the substrate holder; on the surfaces normal to it, n_{nuc} is, in contrast, very low [328]. Likewise, on isolated silicon substrates no increase of the nucleation density is observed [348, 370].

Ion bombardment is thus a necessary (but, as will be shown below, not a sufficient) condition for the enhancement of the nucleation density. With respect to a discussion of the elementary mechanisms of BEN, therefore, the

---

[19] This pressure dependence has recently been confirmed by new investigations by Whitfield et al. [369].

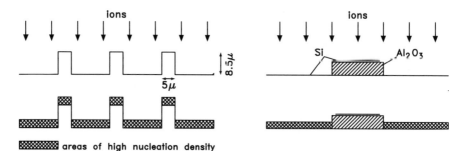

Fig. 4.12. Experiments of Schreck and Stritzker [328] and Jiang et al. [371] (*left*), and McGinnis et al. [370, 348] (*right*) in order to prove the influence of direct ion bombardment during BEN

question of the nature of the ions relevant to BEN (H_x^+ or $C_xH_y^+$) arises, as well as that of their energy distribution (IED) or their average energy.

The bias voltage during a BEN process is typically in the range of 200–300 V. Owing to the working pressures of MW deposition (typically 20–50 mbar) and the resulting mean free path (ca. 0.005 mm [370], see Table 3.5), as well as the typical thickness of the plasma sheath of about 1 mm [367, 370], one has to take into account the fact that the ions undergo a number of collisions in the sheath before they reach the substrate. The average ion energy should thus be considerably lower than the applied bias voltage. Direct measurements of the ion energy distribution, however, are hardly possible owing to the high pressures[20]. McGinnis et al. [370] recently published Monte Carlo simulations for the conditions of BEN. These calculations yield average ion energies of about 15–25 eV and maximum energies of 40–50 eV for bias voltages in the range 200–300 V.

Figure 4.13 shows the dependence of the nucleation density on the bias voltage applied. Below voltages of about 140 V, n_{nuc} decreases drastically. A similar curve for $n_{\text{nuc}}(V_B)$ has recently been published by McGinnis et al. [348]; however, here the voltage required to achieve nucleation densities $> 10^9$ cm^{-2} was on the order of 210 V. McGinnis et al. interpret these observations in terms of a threshold ion energy which is necessary to obtain high nucleation densities. According to the simulations mentioned above, this threshold energy should be about 15 eV.

Nevertheless, a comparison of Figs. 4.13 and 4.9 also allows another interpretation: nucleation densities higher than 10^9 cm^{-2} are achieved only in region III of the I/V curve where the ion current increases significantly (Fig. 4.9: increase of the current at about 140 V). An analysis of the data provided by McGinnis et al. [348] shows that under their conditions this transition occurs at ca. 200 V and thus also correlates well with the first appearance of

---

[20] Sattel et al. [362] postulate from retarding field energy analyzation (RFEA) measurements a mean ion energy of ca. 90 eV. However, as discussed in [344], considerable doubts exist concerning the reliability of these measurements.

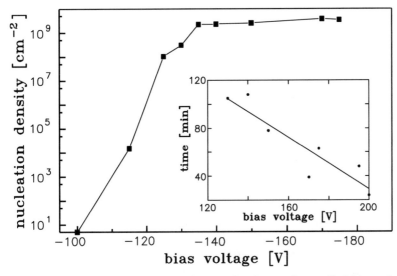

Fig. 4.13. Dependence of the nucleation density on the applied bias voltage. The inset shows the corresponding nucleation times defined by the maximum of the $I(t)$ curve (Fig. 4.10)

high nucleation densities. These observations therefore also allow an interpretation of the results in Fig. 4.13 in terms of a threshold ion dose or a certain ratio of ions and neutral particles which is necessary for nucleation under the conditions of BEN. The fact that the nucleation time decreases with increasing bias voltage, i.e. with increasing ion current (Fig. 4.13), corroborates this speculation.

This situation is a very good example of the complexity of the BEN process emphasized above. Up to now, it has been impossible to control the ion energy and ion current separately. Additionally, owing to the high pressures, measurements of the ion energy distribution have also proven impossible up to now.

The situation is similarly complex with respect to the identification of the nature of the ions relevant to BEN. If carbon-containing ions alone were responsible for the high nucleation densities, one would expect that the density would increase on an increase of the methane concentration in the gas phase (provided the current remains constant and the fraction of carbon-containing species is proportional to the carbon fraction in the gas phase). Unfortunately, the total current decreases with increasing methane content because of a decrease of the plasma density [364] (Fig. 4.14). Nevertheless, a reduction of the nucleation time is observed if the methane fraction is increased from 1 % to 2 % (Fig. 4.14). This is a strong hint of the importance of carbon-containing ions for the BEN process. However, if the methane fraction is further increased to 4 % (which causes a further reduction of the total current), the nucleation time becomes longer again. This may be due to the

150   4. Nucleation Mechanisms

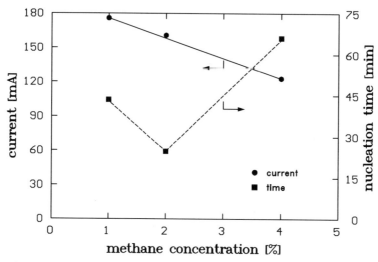

Fig. 4.14. Dependence of the total current at the beginning of the BEN process and of the nucleation time on the methane concentration in the gas phase

fact that hydrogen ions also play a role; but on the other hand it is also imaginable that a certain ion dose or a certain ion/neutral ratio is necessary for BEN. The situation is further complicated by the fact that the composition of the ion current as a function of the methane concentration is not known.

On the basis of these results, it can be excluded that the BEN process relies on H_x^+ ions alone as has been proposed by Yugo et al. [359]; however, it cannot be excluded that hydrogen ions contribute to the process. In this context some experiments of Sattel et al. [372] are of interest which show that plasma beam deposition with 83 eV $C_2H_2^+$ ions does not lead to the formation of diamond nuclei. Obviously, in addition to ion bombardment the presence of hydrogen is a necessary condition; nevertheless, up to now it has not been decided whether the hydrogen has to be in atomic and/or in ionic form.

The dependence of the nucleation density on the substrate temperature is presented in Fig. 4.15, which shows that a minimum temperature of 600°C is required to obtain nucleation densities $n_{nuc} > 10^9$ cm^{-2}. There exist some further data in the literature proving a temperature dependence of n_{nuc}. Nevertheless, these data are not consistent: Gerber et al. [362] report a minimum temperature of 740°C, whereas according to McGinnis et al. [370] high densities can be obtained at 450°C. On the other hand, McGinnis et al. [370], as well as Jiang and Klages [345], report consistently that the nucleation density starts to decrease again above approximately 850°C. Schreck and Stritzker, finally, observed that below 680°C the orientation relationship between the nuclei and the substrate starts to degrade considerably.

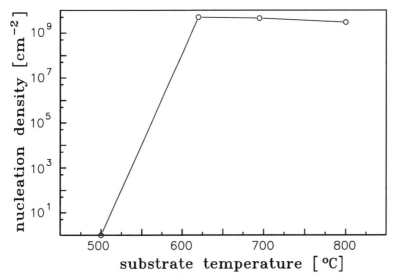

Fig. 4.15. Dependence of the nucleation density on substrate temperature

In any case, an optimum temperature range for BEN seems to exist, even if further investigations are necessary, especially with respect to the minimum temperature required[21]. The decrease of n_{nuc} at very high temperatures is generally ascribed to increased desorption of carbon species and increased hydrogen abstraction [345, 370].

Nevertheless, the temperature dependence of the nucleation density proves beyond doubt that the ion bombardment alone is not sufficient to explain the high nucleation density; otherwise, n_{nuc} should be independent of temperature. Thus, further mechanisms must play a role; since an optimum range of temperature exists, more than one effect must be involved, having opposite influences.

Film Composition: the Nucleation Sequence. In further investigations, the composition of the films deposited during a BEN process has been determined for fixed parameters (2% CH_4, 200 V, 800°C) as a function of time by means of FTIR and AES [344, 346]. Additionally, the stress of these films has been determined (even if only qualitatively [346]). The results of these investigations are summarized in Fig. 4.16 and Table 4.2.

Immediately after starting the BEN process, according to the FTIR measurements, a silicon carbide film forms on the substrate; the thickness of the film first increases with time but reaches a saturation value after about 20 min; by means of the absorption coefficient of SiC it can be estimated to be

---

[21] The inconsistency of the existing data can presumably be attributed in part to the general difficulty of determining the substrate temperature in MWCVD reactors; see, in this context, [106].

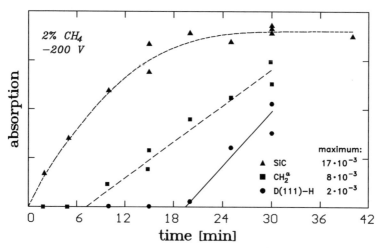

Fig. 4.16. Absorption intensity of the 800 cm^{-1} SiC peak, the 2927 cm^{-1} CH$_2$ peak and the D(111):H peak at 2827 cm^{-1} as a function of time. The intensities are presented in arbitrary units; the maxima of the three curves are given in the figure

on the order of 25 nm [344]. After an incubation time of about 7 min, in addition a polymer-like hydrocarbon layer is observed which is mainly composed of sp^3-bonded CH$_x$ groups (CH$_2$ groups dominate with respect to CH$_3$ and CH groups). This hydrocarbon layer[22] will be labeled a-C:H in the following. After 20 min, finally, the existence of diamond (nuclei) can be inferred from the appearance of the D(111):H peak at 2827 cm^{-1}, which is typical for diamond-containing films [374, 375, 376].

Finally, the fact has to be mentioned that BEN layers possess a high compressive stress on the order of some GPa which appears simultaneously with the D(111):H peak in the spectra of the films.

Every single step of the nucleation sequence presented above has also been observed by other groups; for example, the SiC layer has been described by [365] among others, the a-C:H layer by [368, 373] and, finally, the presence of a large compressive stress by [361, 377]. However, the sequence of their appearance as shown in Fig. 4.16 and Table 4.2 was first established by our group.

Local Characterization. The final step of work that we have performed so far on the bias-enhanced nucleation of diamond consists of an attempt to characterize the films obtained in the first phase of the BEN process on a local

---

[22] The existence of an amorphous carbon layer during BEN has been observed by many authors. However, by far less well established is the fact that this amorphous carbon contains large concentrations of hydrogen. The presence of hydrogen (ca. 30 at%) was recently once again proven beyond doubt by Schreck et al. [373] by means of elastic recoil detection (ERD) with BEN samples which lead to epitaxial growth.

Table 4.2. Nucleation sequence during BEN with 2% CH_4/H_2 at -200 V. —, not observed; o, first appearence; •, fully developed

Time	SiC	a-C:H	Diamond nuclei	Stress
2 min	•	—	—	—
5 min	•	—	—	—
7 min	•	o	—	—
10 min	•	•	—	—
15 min	•	•	—	—
20 min	•	•	o	o
30 min	•	•	•	•

scale by means of field-emission SEM images and AFM measurements, and to correlate the results of these microscopic investigations with the results of the macroscopic characterization presented above [378]. The major findings of this study can be summarized as follows:

- After 7.5 min, isolated particles with a maximum diameter of 20 nm and a maximum height also of 20 nm (i.e. the aspect ratio is one) are found on the substrate surface. Their density is 7×10^{10} cm^{-2}, which is slightly higher than the final density of diamond crystallites.
- These particles have almost completely coalesced after 15 min of BEN; nevertheless, small regions of the Si surface are still uncovered. The particle density has increased slightly at this stage, and, in addition, relatively small clusters are also found. Both facts indicate that further particles have nucleated. The maximum particle dimension is 50×50 nm, and the surface is extremely rough.
- After 21 min the surface is completely covered with a film. The maximum diameter of the particles forming this film is about 100 nm; the roughness has decreased, and the shape of the individual particles appears somewhat rounded. All these observations are typical for a film which has just coalesced from individual islands, which also includes the amalgamation of particles.
- The same trends can also be observed after 27 min: the roughness has decreased further, while the individual particles are even more rounded. Their maximum diameter has only slightly increased, to 110 nm. Finally, it should be noted that no diamond facetes can be observed at this stage.

A similar investigation which was, however, restricted to the first 15 min of the BEN process but decribes these in more detail has been published by Jiang et al. [354, 379]. To a large extent, both studies yield the same results; when correlating these investigations with the macroscopic measurements of the film composition presented above, the following picture emerges.

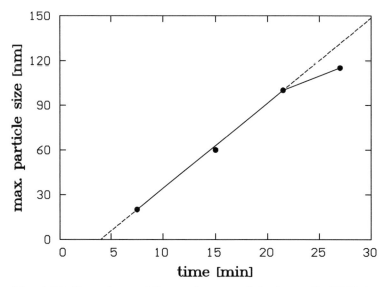

Fig. 4.17. Dependence of the maximum particle size on the BEN time. The incubation time and the time of coalescence are clearly evident

The BEN process is (like the standard nucleation) characterized by an incubation time of about 5–7 min (Fig. 4.17), before the first particles can be observed. During this time, silicon carbide is formed on the substrate surface. Thereafter, particles nucleate with a density slightly exceeding that of the later diamond crystallites. Their first appearance correlates with that of the hydrocarbon layer in the nucleation sequence (cf. Table 4.2); it has therefore to be assumed that these clusters are at least in part composed of a-C:H. The particles possess an aspect ratio of unity; while retaining this aspect ratio, they enlarge until they coalesce. This behavior is typical for the island growth (Volmer–Weber growth) described in Sect. 4.1. Further, a comparison of Figs. 4.16 and 4.17 reveals another correlation between the microscopic morphological observations and the macroscopic nucleation sequence at the time of coalescence (ca. 25 min): as soon as the surface is covered by a continuous hydrocarbon layer the formation of silicon carbide ceases.

After coalescence of the films, only a slight increase of the size of the particles is observed (due to the amalgamation of clusters); the most prominent change with further increasing nucleation time is a pronounced rounding of the individual particles. The most important observation, however, is that even after ca. 30 min no diamond facet can be observed, although according to the nucleation sequence diamond is definitely present in the layers. The appearance of diamond without observable formation of facets at the surface has also been described in [349, 354, 370, 380]; it is generally explained by the existence of small crystallites within an amorphous matrix.

4.3.3 Discussion and Modeling

Role of SiC. Since the invention of BEN, the role of the SiC layer during the BEN process and, especially, in the formation of oriented nuclei has been most controversial. Meanwhile, it has turned out that BEN leads to enhanced nucleation densities and – under certain conditions (see above) – also to heteroepitaxial growth on a variety of substrate materials; in the starting phase of the investigation of BEN, however, the investigations were restricted to silicon and β-SiC. As an example, oriented growth was discovered independently and almost simultaneously by the groups of Stoner and Glass for SiC [353] and of Jiang and Klages [342] for silicon. Even earlier, Stoner et al. [365] had shown that during the BEN process on silicon substrates a SiC layer is created. Thus, a number of authors claimed such a β-SiC layer to be necessary to obtain a high nucleation density and, especially, oriented growth. As a consequence, in order to obtain oriented nucleation on silicon Wolter et al. [381] introduced a carburization step prior to the standard BEN process, which has been copied several times by other researchers [332, 340, 352]. In order to explain the necessity for this β-SiC layer these authors used arguments based on the better lattice-matching (silicon/diamond: 54 %; β-SiC/diamond: 22 %; see Table 4.1) as well as the creation of a diffusion barrier [339] leading to a carbon saturation on the surface.

The work of Milne et al. [352], as well as the investigations of Jiang and Klages [342] which have already been mentioned, however, show that this carburization step – while indeed reducing the necessary BEN time – is not absolutely necessary to obtain either high nucleation densities or oriented nucleation. Nevertheless, this result is still in agreement with the formation of a SiC layer during the BEN process itself; new TEM investigations [338, 355, 382], however, of the interface between the silicon substrate and oriented diamond nuclei prove beyond doubt that direct nucleation of diamond on silicon is possible. These TEM images show a direct 2:3 lattice-matching without any interlayer. On the same samples crystallites can also be found below which an amorphous SiC layer is present; however, the orientation of these crystallites with respect to the substrate is rather poor.

Taking into account the results of our own investigations presented above, the following conclusions can be drawn. The formation of SiC, which is observed during all diamond deposition processes, is promoted by the conditions of BEN[23]. It occurs automatically and can only be avoided if the substrate surface is covered immediately after the start of the BEN by diamond crystallites or their precursor (see, in this context, also [371]). This conclusion is in agreement with the fact that the formation of SiC stops as soon as the substrate is completely covered (see above). Whether this layer is necessary for the nucleation, for example serving as a diffusion barrier, is not clear at the present stage. It is, however, clear that it is suited for epitaxial nucleation

---

[23] This is evident from some of our own investigations on the thickness of the SiC layer with and without bias voltage, keeping all other parameters constant [346].

only if it consists of oriented, crystalline β-SiC. Other modifications such as amorphous SiC, while still allowing high nucleation densities, lead inevitably to randomly oriented diamond nuclei only. Strategies to obtain films as well oriented as possible should thus either include the deposition of epitaxial β-SiC-layers or attempt to avoid any SiC interlayer by immediate diamond nucleation.

Role of the Hydrocarbon Layer. Very similarly to the role of silicon carbide, the role of the hydrocarbon layer initially formed on the substrate is also to a great extent still unclear; moreover, in many of the (in most cases only speculative) attempts to explain the BEN existing up to now in the literature, this layer is not taken into account. The existence of this layer, however, is beyond doubt, as is evident from the above presentation. This is true for both the oriented and the random nucleation.

Concerning this point, there are some similarities between the BEN, on the one hand, and the deposition of ta-C and c-BN, on the other hand, which also starts with the formation of sp^2-bonded material (cf. Sects. 4.4 and 4.5). Indeed, Robertson [360], for example, tried to describe common mechanisms for all three processes, making use of the fact that each of them relies on ion bombardment. He proposed that – as in the case of c-BN discussed below – the ion bombardment induced the formation of a texture of the sp^2 layer with the graphite-like layers standing perdendicular on the substrate, owing to the compressive stress created. This texture then leads to the nucleation of diamond via a 2:3 lattice-matching as is indeed the case for c-BN and – as we have seen – for the nucleation on HOPG. However, such textured sp^2 structures have not been observed up to now; in addition, the formation of the high compressive stress takes place simultaneously with, not prior to, the nucleation of diamond (Table 4.2). Finally – and this is the strongest argument against the model of Robertson – in this case the growing diamond layer should possess a $\langle 111 \rangle$ texture (compare Figs. 4.7 and 4.19). In contrast, it is unequivocally evident from the existing data that the orientation of the growing film reflects that of the substrate [371], as is for example the case for the TEM images of the interface between the substrate and diamond nuclei discussed above.

Thus the role of the a-C:H layer during BEN is still open to discussion. One possibility is that it just serves as a carbon reservoir (supersaturation). One can also imagine that this layer contains graphitic or aromatic precursors which — according to the considerations of Angus et al. [320] discussed above — serve as starting points for the formation of diamond nuclei. Proofs of this and other hypotheses are unfortunately rendered difficult by the fact that the a-C:H layer is etched back to a large extent by atomic hydrogen after switching from BEN to growth conditions.

Stress. Table 4.2 shows that, simultaneously with the first diamond particles, a large compressive stress is observed in the BEN layers, which is – as in the case of c-BN deposition – caused by the ion bombardment. Gerber et al. [377] as well as – as we have seen – Robertson [360] developed models which assume this stress to be the reason for the formation of diamond nuclei. However, as in the case of c-BN, the large compressive stress may be not the reason for but, rather, the consequence of the nucleation of diamond[24] [72, 214]. Defects created by the ion bombardment lead to a volume change dV/V and this in turn leads, via

$$\sigma = \frac{\mathcal{E}}{2(1-2\nu)} \frac{dV}{V} , \qquad (4.19)$$

to a compressive stress (cf. (3.46)). Here, $\mathcal{E}$ is the Young's modulus and ν Poisson's ratio. The Young's modulus of diamond of 1140 GPa (Table 2.1) is the highest of all materials, whereas for hydrocarbon layers $\mathcal{E}$ can be estimated as 200 GPa at the highest. Therefore, a sharp increase of the stress with the onset of the formation of diamond in the layers can be explained rather easily.

It has to be taken into account further that the occurrence of stress roughly correlates with the coalescence of the films, as is evident from a comparison of Table 4.2 and Fig. 4.17. Since a continuous film is a prerequisite for the formation of stress, all those models must be excluded which rely on the effects of stress prior to the coalescence of the individual crystals.

Mechanisms of Diamond Nucleation. BEN has two positive effects: on the one hand it yields a drastic increase of the nucleation density, and on the other hand it leads under certain conditions to the formation of oriented nuclei. Any model of BEN must therefore be able to explain both effects. However, both effects need not necessarily be due to the same mechanisms. It is an obvious fact that the parameter ranges required to obtain high n_{nuc} and oriented nuclei are not identical; otherwise, all BEN experiments would yield not only high densities of nuclei but also oriented nuclei. This, however, is not the case, as is also evident from our own investigations presented above. A possible explanation is the deterioration of originally oriented growth due to the ion bombardment in our experiments, since it has already been known for some time that excessively long BEN times can lead to a loss of orientation[25]. But, on the one hand, Schreck et al. [383] observed oriented nuclei with relatively (!) low densities for very short bias times; on the other hand, Milne et al. [352], for example, report that very short bias times result in high nucleation densities but do not lead to oriented nuclei. Evidently, the dependences of the nucleation density and the orientation of the nuclei on the process parameters (bias voltage, bias time, pressure, substrate temperature, gas phase composition) are different; this, however, means also that the underlying mechanisms are different.

---

[24] Compare in this context Sect. 3.3.5 and especially (3.46).
[25] Compare also the discussion of the influence of ion bombardment on the quality of diamond crystallites in Sect. 3.3.6.

The mechanisms for the drastic increase of the nucleation density have not been explained satisfactorily up to now. Nevertheless, at this point at least the following fact should be emphasized which has to be taken into account in any future modeling of BEN: in order to obtain high nucleation densities, both ion bombardment (see Fig. 4.12) and the presence of atomic hydrogen (compare the plasma beam experiments of Sattel et al. [372] discussed above) are prerequisites sine qua non. Thus all models of BEN must include a physical as well as a chemical component. Chemical mechanisms alone, on the one hand, and ion-induced effects alone, on the other hand, cannot explain the BEN.

BEN and Heteroepitaxy. Finally, the mechanisms leading to the formation of oriented nuclei have to be addressed. As with all aspects of BEN, in this case also the state of knowledge is far from a complete understanding. Nevertheless, some new experiments provide further insight into the relevant processes; these experiments will therefore be discussed very briefly:

- Jiang et al. [248] investigated the growth of diamond on a pregrown thick diamond layer with random crystallites under the conditions of BEN. After a certain time, a $\langle 100 \rangle$ texture is observed within the films. However, the individual, relatively small crystallites (maximum 0.5 μm) are oriented with respect to each other only within regions of a few microns in size. The resulting layer possesses a compressive stress and a quality which is inferior to that of the substrate film.
- On a large crystal with a $\{111\}$ surface parallel to the substrate surface, no homoepitaxial growth is observed under BEN conditions. Rather, a $\langle 100 \rangle$-textured layer consisting of small crystallites is observed, which has obviously been formed by renucleation.
- On $\{100\}$ surfaces parallel to the substrate surface, the epitaxial growth continues. But here also renucleation is observed: the crystallites become smaller and some possess a misorientation of a few degrees.

This study, as well as similar experiments presented in [384], shows that ion bombardment obviously has the effect that the diamond crystallites orient themselves in such a manner that one [100] axis is in the direction of the bombardment[26]. This could, for example, be caused by a channeling effect or the selective sputtering of different crystal faces, as has been observed in other systems [385]. A second important observation is the fact that secondary nucleation is a mechanism relevant to a change of texture. On the other hand, the experiments of Jiang et al. show that information about the orientation of the underlying substrate also plays a role. Taking all this together, this means that the orientation of the freshly nucleated species is determined by the ion bombardment as well as by the substrate.

---

[26] A really striking proof of this assumption might be obtained from experiments using an ion bombardment which is not normal to the substrate. This, however, can hardly be realized under the conditions of diamond deposition.

It has to be mentioned, finally, that the ion bombardment is at least in part responsible also for the creation of the misorientation. Experimental evidence for this conclusion can be found in the work of Jiang et al. [248, 384] and especially in a recent paper by Schreck et al. [383].

Conclusions. The facts concerning the influence of the ion bombardment during BEN presented so far can be summarized as follows:

- Ion bombardment leads to a high nucleation density on a variety of substrates.
- Under the influence of the ion bombardment the crystals orient themselves in such a manner that one $\langle 100 \rangle$ axis points in the direction of the bombardment.
- The crystallites retain, at least in part, information about the orientation of the substrate.
- The ion bombardment leads, at least under certain conditions, to an azimuthal misorientation.

The ion bombardment during BEN causes – for reasons which are still not understood, as discussed above – a high nucleation density, for which, nevertheless, the presence of atomic hydrogen is a second necessary condition. These nuclei adopt an orientation which is influenced by two factors: at least after prolonged ion bombardment, one $\langle 100 \rangle$ direction is aligned parallel to the bombardment; on the other hand, information about the substrate orientation still remains, at least in part. It is therefore imaginable that diamond nuclei which have been formed at the beginning of the BEN process and which are thus in direct contact with the substrate orient one $\langle 100 \rangle$ direction normal to the substrate because of the ion bombardment, but all other directions according to the substrate since they are – owing to the high temperatures and, eventually, also the energy transferred to the substrate by the ion bombardment – relatively mobile, and since in every case the epitaxial orientation is energetically favorable. But misorientations are introduced, also because of the ion bombardment, and – if the bombardment is prolonged or too strong – renucleation takes place, leading finally to a loss of the orientation.

4.4 Nucleation of c-BN

The sputter model presented in Sect. 3.3.5 has been shown to be suitable to describe adequately all experimental observations published up to now concerning the growth of c-BN (with the exception, however, of the experiments discussed in Sect. 3.3.5); nevertheless, the mechanisms postulated require the existence of c-BN nuclei. The formation of such nuclei, i.e. the nucleation of c-BN, is not addressed by the model (at least in the first instance).

4.4.1 Experimental Observations

Experimental investigations of the nucleation of c-BN have been scarce for a long time. In 1993, Kester et al. [386] published a TEM image of the interface between a silicon substrate and c-BN film which gave clear evidence that c-BN is not formed directly on the substrate but that there exists a well-defined sequence of various layers:

a-BN $\longrightarrow$ textured h-BN (c axis $\parallel$ substrate) $\longrightarrow$ c-BN.

The amorphous layer is only a few nanometers thick and has not yet been characterized with respect to its composition. The subsequent h-BN layer possesses a (0002) texture, which means that the c axis lies parallel to the substrate while the BN six-membered rings are oriented normal to the substrate; concerning the ab planes, however, no preferential orientation exists. The c-BN layer itself, finally, is nanocrystalline. In Fig. 4.18, showing a TEM image of a c-BN layer deposited by means of mass-separated ion beam deposition (MSIBD)[27], this sequence of layers is clearly discernible.

This characteristic nucleation sequence has, meanwhile, been confirmed many times. It is observed irrespective of the method (PVD or CVD)[28]; furthermore, it is found not only on silicon substrates but also on SiC [388], diamond [389], textured h-BN with the c axis normal to the substrate [203] and silicon substrates with various adhesion layers [390]. It has to be pointed out, however, that in quite a few cases, which will be discussed in more detail below, c-BN films have been grown directly on the substrate without an intermediate textured h-BN layer [239, 391].

Figure 4.19 shows an enlarged section of the TEM image of Fig. 4.18, which provides some further information which is of relevance to the explanation of the nucleation of c-BN. There is a crystallographic relationship between the textured h-BN and the c-BN crystallites formed on top: The h-BN (0002) planes are parallel to the c-BN (111) planes. Thereby, three c-BN (111) planes (spacing 0.21 nm) correspond to two h-BN (0002) planes (spacing 0.31 nm), which means that (as in the case of the nucleation of diamond on HOPG, see Sect. 4.2.3) a 2:3 lattice-matching occurs.

The same crystallographic relationships as those existing between the h-BN nucleation layer and the c-BN crystallites as shown in Fig. 4.19 have also been observed for the 1–2 nm thick h-BN material which is usually present at the grain boundaries between c-BN crystallites [392]. This holds for HPHT crystallites and PVD films.

Before discussing the mechanisms leading to the formation of this nucleation sequence and thus finally to the nucleation of c-BN, some further observations which are of relevance to this discussion are listed in the following:

---

[27] This layer was prepared in the framework of a cooperation with the University of Konstanz [219].
[28] The existence of this nucleation sequence with ion-assisted CVD methods was first proven by our group in [387].

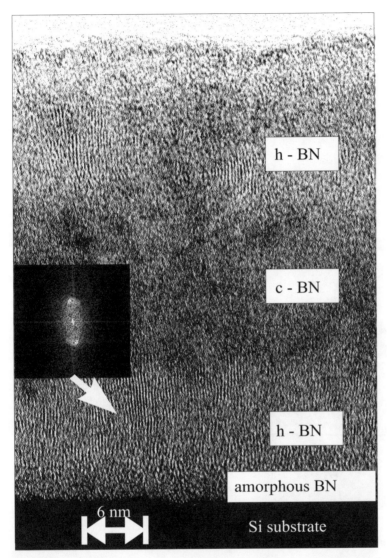

Fig. 4.18. TEM cross section of a c-BN film deposited with MSIBD [219]

1. The thickness of the textured h-BN layer varies considerably depending on the deposition method and parameters; the range of values reported in the literature stretches from 10 to 100 nm. Systematic studies concerning this aspect are still missing; nevertheless there seems to be a tendency for it to be higher for CVD methods and plasma-assisted PVD methods than for ion beam techniques [72, 202, 204].

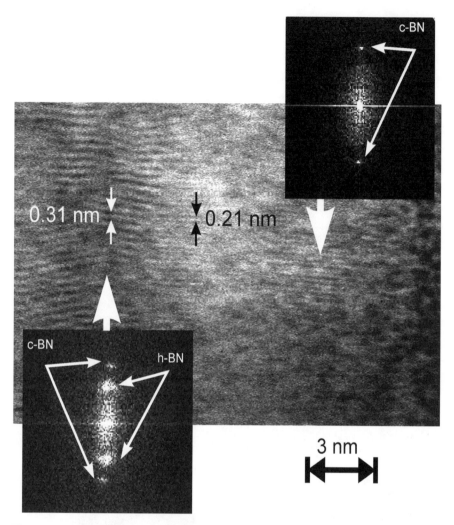

Fig. 4.19. High-resolution TEM image of the transition from h-BN (*left*) to c-BN (*right*). In addition, diffraction patterns of the respective regions are presented

2. The interface between the textured h-BN layer and the subsequent c-BN layer is not smooth, but shows a roughness in the nanometer range. This is clearly evident in Fig. 4.18.
3. In Fig. 4.18, above the first nanocrystalline c-BN layer another textured h-BN layer can be distinguished, which is again followed by c-BN. To explain this sequence it has to be mentioned that the MSIBD deposition was interrupted after some time for 72 hours, during which time the substrate temperature was lowered from 350°C to room temperature; nevertheless, the sample remained in UHV during this period [219]. Ob-

viously, owing to this interruption of the growth process a new nucleation step was necessary.
4. Recent results of Röder et al. [393] prove that the formation of the (0002) texture alone is not sufficient for the nucleation of c-BN. Transmission electron microscope images show that the formation of the texture starts to develop at bias voltages far below the voltages required for the formation of c-BN.
5. The results presented in Sect. 3.3.5 show, on the other hand, that nucleation of c-BN is possible only within the c-BN region of the $F/E_\mathrm{i}/T_\mathrm{s}$ parameter space (at least for indirect energy input).

4.4.2 Mechanisms of Nucleation

On the basis of the experimental observations presented above, it is possible to formulate a model of the nucleation which is suited to explaining these results at least quantitatively.

First, the geometrical relationships are illustrated again in Fig. 4.20. In the h-BN (0002) plane as well as in the c-BN (111) plane the atoms are arranged in hexagons; however, the c-BN (111) plane is not flat but, rather, buckled. Nevertheless, in the projection the distances between the atoms are nearly identical. These geometrical relationships, as well as the observed crystallographic relationships between the h-BN nucleation layer and the c-BN crystallites, suggest that – as in the case of the nucleation of diamond on HOPG – a 2:3 lattice-matching plays an important role (compare Table 4.1). The mechanisms leading to this matching, however, have to be different for diamond and c-BN, despite all geometrical analogies. In the case of diamond atomic hydrogen plays a decisive role, as has been discussed in Sect. 4.2.3; in the case of c-BN the nucleation is, rather, dominated by the ion bombardment, as is evident from the above listing of experimental results.

It is generally agreed that the amorphous layer inevitably observed at the interface between the substrate and BN layer is formed by 'ion mixing' due to the ion bombardment. Its thickness corresponds roughly to that to the ion range; its composition is $\mathrm{Si}_x\mathrm{B}_y\mathrm{N}_z$.

After the formation of this mixed layer, according to the model of competitive growth, first h-BN starts to grow owing to the lack of c-BN crystallites[29]. Although the nucleation step takes place in the c-BN region of the F/E_i parameter space, according to the model h-BN is stable at least up to $v_h = 0$; in addition, the stability is probably enhanced by the formation of the (0002) texture (reduction of sputter processes due to channeling effects) [72].

The (0002) texture of this layer is due, as is again generally agreed, to the large biaxial stress which is caused by the intensive ion bombardment. Calculations of McKenzie et al. [232] show that this texture provides the

---

[29] Since the a, b axes of this layer possess a rather random distribution, this material should correctly be labeled turbostratic (t-BN) rather than hexagonal BN.

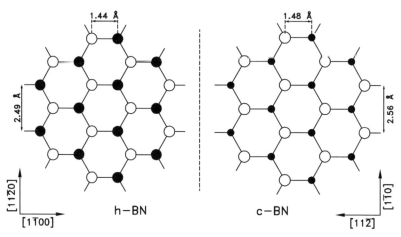

Fig. 4.20. Schematic representation of the (0001) plane of h-BN and the (111) plane of c-BN. The latter is not flat but 'buckled'. The diagram represents a projection; atoms below the plane are indicated by a smaller radius. With slightly different distances (1.42/1.45 Å and 2.46/2.52 Å), this diagram is also valid for graphite and diamond (according to [330, 392])

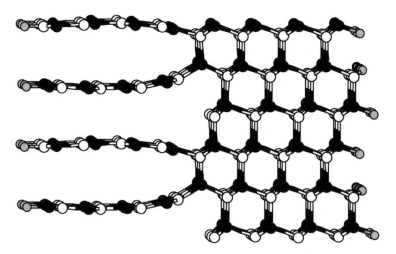

Fig. 4.21. Structure of the interface between a textured h-BN and a c-BN layer calculated by means of molecular-dynamic simulations [98]. Open symbols, nitrogen; black symbols, boron; grey symbols, hydrogen

thermodynamically most favorable arrangement in the presence of a biaxial stress.

Whereas there is agreement in the literature concerning these first two steps of the nucleation process, the formation of the first c-BN crystallites on the textured h-BN is still controversial. From the above compilation of experimental observations concerning the nucleation of c-BN, it is evident that the mere existence of the textured h-BN layer is not sufficient for the nucleation of c-BN, which, rather, takes place only if the ion bombardment is strong enough. At least in the case of indirect energy input, nucleation is possible within the c-BN region of the F/E_i diagram.

At this point it is necessary to remember once again the geometrical analogies between h-BN and c-BN (Fig. 4.20) as well as the experimentally observed crystallographic relationship between the nucleation layer and the c-BN layer. These suggest a nucleation via 2:3 lattice-matching in analogy to the nucleation of diamond on HOPG. Figure 4.21 shows a model of the transition from h-BN to c-BN obtained from molecular-dynamic calculations by Widany et al. [98]. Distortions of the crystal lattice are limited to a narrow region of 0.35 nm thickness; energetic considerations show this configuration to be stable [98][30].

In the framework of our extended model of c-BN deposition (the model of competitive crystal growth) the formation of c-BN crystallites on the textured h-BN nucleation layer can be explained as follows [97]. Supported by the ion bombardment, material (especially boron atoms) is introduced between the individual sheets of the hexagonal BN, thus causing densification. It can be seen from Fig. 4.21 that in this case the formation of c-BN can take place by slight changes of the bond lengths and formation of new bonds without breaking existing bonds and without diffusion processes. These considerations are in agreement with the finding in Sect. 3.3.5 that nucleation of c-BN is only possible under conditions allowing an effective densification (cf. (3.53) and Fig. 3.25).

Taking all together, the above discussion of the nucleation of c-BN and the model of competitive crystal growth presented in Sect. 3.3.5 form a comprehensive picture of the nucleation and growth of c-BN films, which is shown schematically in Fig. 4.22.

First an amorphous $Si_xB_yN_z$ layer is formed by ion mixing. As soon as the thickness of this layer exceeds the ion range, hexagonal BN starts to grow. The strong ion bombardment creates a large biaxial stress, which leads to the formation of the energetically favorable (0002) texture. By introduction of material between the individual sheets of the hexagonal nucleation layer, local densification takes place, leading to the nucleation of c-BN crystallites by means of 2:3 lattice-matching. Since the growth velocity of c-BN is higher than that of h-BN owing to selective sputtering, the h-BN crystallites are gradually overgrown and vanish. This explains also the relatively rough interface between the nucleation layer and c-BN film (Fig. 4.18). By secondary nucleation (keeping in mind the small crystallite size of the c-BN films), finally, the original (111) texture is lost.

While the description discussed above considers the nucleation and growth of c-BN as separate processes, which nevertheless do not contradict each other

---

[30] This result is supported by the observation of textured h-BN at the grain boundaries in c-BN films as well as in HPHT c-BN.

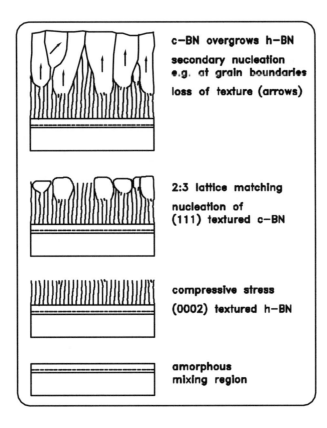

Fig. 4.22. Schematic summary of c-BN nucleation and growth

but, rather, form a comprehensive picture, nucleation is hardly taken into account in the other models of c-BN deposition which have been discussed in Sect. 3.3.4. All those descriptions suggesting a transition $sp^2 \longrightarrow sp^3$ in the interior of the films, such as the stress models and the model of direct subplantation (Table 3.6), are based on the assumption that nucleation in the true sense of the word does not take place. The quenching model, in the formulation of Hofsäss and Ronning [26], and the model of indirect subplantation of Robertson [223], on the other hand, presume – like the sputter model – the existence of c-BN nuclei; concerning their formation, however, only some general statements (e.g. 2:3 lattice-matching) are given but no explicit description has been developed.

At the end of this section on the nucleation of c-BN, an aspect of technological relevance should be mentioned briefly; for reasons of space, we nevertheless refrain from a detailed presentation.

It has been stated in Sect. 3.3.1 that the application of c-BN films is, at the present time, prevented mainly by their poor adhesion. This is due not only to the high compressive stress of the layers but, to a large extent, also their low adhesion strength[31]. The adhesion strength of thin films is determined by the interaction between the film and substrate at the interface. A thorough analysis of the complex substrate/nucleation layer/c-BN interface [214, 215, 217] revealed that, from a mechanical point of view, the textured h-BN nucleation layer is the weakest link of the system. Besides conventional measures to improve the adhesion strength which are well-known from other thin-film systems (e.g. adhesion interlayers, rough interfaces), a solution of the adhesion problem may therefore also be found in the avoidance of this nucleation layer. It has already been mentioned that there are some hints in the literature that this might indeed be possible [239, 391]. In these experiments, boron-rich interlayers were used. Since, however, only a small amount of data exists, it would be too early to generalize these tendencies or to speculate on possible nucleation mechanisms in the absence of the textured layer[32].

Summarizing, it can therefore be stated that in order to solve the stress/adhesion problem, both the growth and the nucleation of c-BN have to be optimized: the growth process determines the stress of the films, as has been shown in Sect. 3.3.5; their adhesion strength, on the other hand, is mainly influenced by the nucleation step.

4.5 Nucleation of ta–C and β-C_3N_4

It has been shown in Sect. 3.4 that the formation of ta-C takes place via direct or indirect subplantation of carbon atoms in near-surface regions, i.e. in the interior of an sp^2-bonded carbon layer. A 'nucleation' of ta-C in the true sense of the word, i.e. formation of ta-C nuclei on a foreign substrate, therefore does not take place. Although the deposition of the original sp^2 layer also requires a nucleation step, this step is without relevance to the actual formation of ta-C.

In this respect, the deposition of ta-C differs distinctly from diamond, as well as from c-BN deposition, for each of which a nucleation step is observed which is clearly distinguished from the subsequent growth process.

Up to now, the deposition of crystalline β-C_3N_4 films has not been achieved beyond doubt, as has been shown in Sect. 3.5; as a consequence it is not possible to discuss possible mechanisms of nucleation.

For this reason, at this place only one observation of Marton et al. [71] should be mentioned briefly. In contrast to the deposition of ta-C, no phase evolution is observed during the deposition of amorphous $C_{1-x}N_x$ layers containing a particular fraction of tetrahedrally bonded carbon (Sect. 3.5). This means that the tetrahedral bonds are not created by a subsequent phase

---

[31] High intrinsic stresses of course cause peeling of the films in the case of low adhesion strengths. This is, however, also true for external mechanical stresses.

[32] Some of our own preliminary experiments show that it is at least possible to reduce the thickness of the nucleation layer and that this reduction already leads indeed to an improvement of the adhesion strength [216, 217].

transition $sp^2 \longrightarrow sp^3$ but must, rather, form directly during the deposition of the atoms at the surface or slightly below. To this extent, a nucleation step may indeed play a role during the deposition of partly tetrahedral $C_{1-x}N_x$ layers.

4.6 Conclusions

The discussion in the preceding sections has shown that the mechanisms of nucleation of diamond-like superhard materials are, by far, not as well understood as the growth mechanisms. In addition it has become evident that there are hardly any common aspects between the materials diamond, c-BN and ta-C; moreover, in the case of diamond several processes exist which may lead to the formation of nuclei. Nevertheless, some important trends can be identified which will be briefly presented in the following; the major mechanisms are summarized in Table 4.3.

In the first instance it has to be pointed out that in all cases classical nucleation phenomena, as discussed in Sect. 4.1, do not play a role (with the exception of diamond nucleation at defects)[33]. Rather, in almost all cases a modification of the nucleation surface takes place first, either by chemical reaction (carburization) or by the deposition of layers (a-C:H, h-BN) the properties of which are distinctly different from those of the superhard material to be deposited; these layers are subsequently transformed into the diamond-like modification by mechanisms which differ from case to case and which are not completely understood at present. It is not even clear in all cases to what extent these surface modifications are necessary for the nucleation process. The carburization during diamond deposition creates a diffusion barrier, which finally leads to carbon supersaturation at the surface; during BEN, however, carburization is not absolutely necessary, according to the discussion in Sect. 4.3.3. Likewise, it seems (even if this fact has not yet been proven finally) that under certain conditions c-BN nucleation is possible without the nucleation sequence described in Sect. 4.4 taking place. In the case of ta-C, on the other hand, the creation of sp^3 bonds occurs below the surface within an originally sp^2-bonded layer which is therefore absolutely necessary for the formation of ta-C. However, here one can hardly speak of a nucleation process in the true sense of the word.

When comparing the nucleation mechanisms of diamond and c-BN, two analogies seem to be conspicous at first sight: first, 2:3 lattice-matching plays a role during the nucleation of diamond on HOPG as well as during the c-BN nucleation sequence; in addition, both the BEN process and the nucleation of c-BN are ion-induced. On closer examination, however, these analogies do not

---

[33] However, during the actual growth of diamond and also in the case of homoepitaxy, the phenomena decribed in Sect. 4.1 are indeed involved in the formation of new monolayers.

4.6 Conclusions 169

Table 4.3. Summary of the nucleation mechanisms (NM) of diamond-like superhard materials (on silicon). Listed in the table are the mechanisms (as far as they are known) and the surface modifications (SM) which play a role in the processes. For diamond and c-BN, the various different cases (C) discussed in detail in the this chapter are listed

	Diamond	
C	Untreated substrates	Abrasive treatment
NM	Nucleation on defects	Homoepitaxy
SM	Carburization	Carburization

[illustration: left panel shows a diamond crystal on carburized layer with substrate surface; right panel shows diamond debris embedded in carburized layer]

	Diamond	
C	HOPG	BEN
NM	2:3 lattice-matching	???
	Chemically induced	
SM	—	Carburization
		a-C:H layer

[illustration: left panel shows diamond on HOPG; right panel shows a-C:H layer with substrate surface and carburized layer]

Table 4.3. continued

	c-BN	
C	Standard process	Boron-rich surfaces
NM	2:3 lattice-matching Physically induced (densification)	Densification
SM	Ion-mixed layer Textured h-BN	Ion-mixed layer
	ta-C	C_3N_4
NM	Phase transition	???
SM	sp^2 carbon layer	???

reach far. The 2:3 lattice-matching is caused by the geometrical similarities of the materials involved; the processes leading to these matchings are nevertheless quite distinct: for diamond nucleation on HOPG, chemical reactions driven by atomic hydrogen are decisive, for c-BN nucleation, on the other hand, local densification caused by ion bombardment is decisive. Likewise, the ion-induced processes relevant during BEN are very different from those determining c-BN nucleation. Although these processes have yet not been identified for BEN, the differences are evident from the different parameter ranges as well as from the very fact that the presence of atomic hydrogen is a sine qua non in the case of BEN. Chemical mechanisms therefore have to be involved which are of no importance for c-BN nucleation.

Finally it seems that – like the growth – the nucleation of diamond also relies on chemical mechanisms, even if a physical component is also involved during BEN, whereas the nucleation of c-BN is caused by physical, ion-induced processes.

Both materials nevertheless have in common the fact that the nucleation is by far more difficult to achieve than the growth. The formation of the first diamond and c-BN nuclei can only be obtained under special conditions, wherease the growth – at least in a certain parameter space – takes place rather spontaneously by attachment processes which are either chemically or physically stimulated, as soon as the first nuclei have been formed. The reasons for these difficulties have not been identified so far; however, it can be taken for certain that the high surface energies, which are common to all diamond-like superhard materials, play an important role.

References

1. R. Kassing, E. Oesterschulze: "Sensors for scanning probe microscopy", in *Micro/Nanotribology and Its Application*, ed. by B. Bhushan (Kluwer, Amsterdam, 1997), NATO ASI Series Volume 330, p. 35
2. M.L. Cohen: Solid-State Commun. **92**, 45 (1994)
3. P.S. Kisly: Inst. Phys. Conf. Ser. **75**, 107 (1986)
4. J.N. Plendl, P.J. Gielisse: Phys. Rev. **125**, 828 (1962)
5. R.J. Goble, S.D. Scott: Canadian Mineralogist **23**, 273 (1985)
6. M.Y. Chen, X. Lin, V.P. Dravid, Y.W. Chung, M.S. Wong, W.D. Sproul: Surf. Coat. Technol. **54/55**, 360 (1992)
7. J.C. Anderson, K.D. Leaver, R.D. Rawlings, J.M. Alexander: *Material Science* (Chapman & Hall, London, 1990), Chap. 9, Mechanical Properties, p. 181
8. A. Kelly, N.H. MacMillan: *Strong Solids* (Clarendon, Oxford, 1986)
9. L. Bergmann, C. Schäfer: *Lehrbuch der Experimentalphysik*, Vol. I, Mechanik Akustik Wärme (Walter de Gruyter, Berlin, 1974), Chap. 5, Elastizität der festen Körper, p. 230
10. W.D. Callister: *Material Science and Engineering* (Wiley, New York, 1994)
11. H. Frey, G. Kienel: *Dünnschichttechnologie* (vdi Verlag, Düsseldorf, 1986)
12. D.R. Askeland: *The Science and engineering of materials* (Chapman & Hall, London, 1996)
13. C. Brookes: Inst. Phys. Conf. Ser. **75**, 207 (1986)
14. C. Friedrich, G. Berg, E. Broszeit, C. Berger: Mat.–wiss u. Werkstofftech. **28**, 59 (1997)
15. T.F. Page, S.V. Hainsworth: Surf. Coat. Technol. **61**, 201 (1993)
16. S. Veprek, S. Reiprich, L. Shizhi: Appl. Phys. Lett. **66**, 2640 (1995)
17. P. Calvert: Nature **357**, 365 (1992)
18. H. Holleck: J. Vac. Sci. Technol. A **4**, 2661 (1986)
19. P.B. Mirkarimi, L. Hultman, S.A. Barnett: Appl. Phys. Lett. **57**, 2654 (1990)
20. K.K. Shih, D.B. Dove: Appl. Phys. Lett. **61**, 654 (1992)
21. M.L. Cohen: Phys. Rev. B **32**, 7988 (1985)
22. A.Y. Liu, M.L. Cohen: Science **245**, 841 (1989)
23. J.C. Anderson, K.D. Leaver, R.D. Rawlings, J.M. Alexander: *Material Science* (Chapman & Hall, London, 1990), Chap. 8, Crystal Defects, p. 151
24. W. Harrison: *Electronic Structure and the Properties of Solids* (W.H. Freeman, San Francisco, 1980)
25. L. Pauling: *The nature of Chemical Bonds and the Structure of Molecules and Crystals* (Cornell University Press, Ithaca, N.Y., 1960)
26. H. Hofsäss, C. Ronning: "Ion beam deposition and doping of diamondlike materials", in *Beam Processing of Advanced Materials*, ed. by J. Singh, S. Copley, J. Mazumder (Conference Proceedings of ASM, 1996), p. 29
27. D.M. Teter, R.J. Hemley: Science **271**, 53 (1996)
28. D.M. Teter: MRS Bulletin p. 23 (January 1998)

29. M.L. Cohen: Mat. Sci. Eng. A **209**, 1 (1996)
30. H. Ehrhardt: Surf. Coat. Technol. **74/75**, 29 (1995)
31. A.Y. Liu, M.L. Cohen: Phys. Rev. B **41**, 10 727 (1990)
32. A.Y. Liu, R.M. Wentzcovitch: Phys. Rev. B **50**, 10 362 (1994)
33. R. Messier, A.R. Badzian, T. Badzian, K.E. Spear, P. Bachmann, R. Roy: Thin Solid Films **153**, 1 (1987)
34. A.Y. Liu, R.M. Wentzcovitch, M.L. Cohen: Phys. Rev. B **47**, 1760 (1989)
35. M. Kawaguchi, T. Kawashima: J. Chem. Soc., Chem. Commun. p. 1133 (1993)
36. S. Nakano, M. Akaishi, T. Sasaki, S. Yamaoka: Mat. Sci. Eng. **A 209**, 26 (1996)
37. M.O. Watanabe, T. Sasaki, S. Itoh, K. Mizushima: Thin Solid Films **281/282**, 334 (1996)
38. M.L. Cohen: Mat. Res. Soc. Symp. Proc. **383**, 33 (1995)
39. A. Hollemann, E. Wiberg: *Lehrbuch der Anorganischen Chemie* (Walter de Gruyter, Berlin, 1985), Chap. XVII.1.6., Stickstoffverbindungen des Bors
40. M. Cohen: Science **234**, 549 (1986)
41. H.W. Kroto, J.R. Heath, S.C. O'Brien, R.F. Curl, R.E. Smalley: Nature **318**, 162 (1985)
42. W. Krätschmer, L.D. Lamb, K. Fostiropoulos, D.R. Huffman: Nature **347**, 354 (1990)
43. W. Krätschmer: Phys. Bl. **48**, 553 (1992)
44. J. Fink, E. Sohmen: Phys. Bl. **48**, 11 (1992)
45. S. Iijima: Nature **354**, 56 (1991)
46. T.W. Ebbesen, P.M. Ajayan: Nature **358**, 220 (1992)
47. Z. Weng-Sieh, K. Cherrey, N.G. Chopra, X. Blase, Y. Miyamoto, A. Rubio, M.L. Cohen, S.G. Louie, A. Zettl, R. Gronsky: Phys. Rev. B **51**, 11 129 (1995)
48. A.Y. Liu, M.L. Cohen, K.C. Hass, M.A. Tamor: Phys. Rev. B **43**, 6742 (1991)
49. R. Riedel: Adv. Mater. **4**, 759 (1992)
50. S. Ulrich, T. Theel, J. Schwan, H. Ehrhardt: Surf. Coat. Technol. **97**, 45 (1998)
51. H. Spicka, M. Griesser, H. Hutter, M. Grasserbauer, S. Bohr, R. Haubner, B. Lux: Diamond Relat. Mater. **5**, 383 (1996)
52. A. Bartl, S. Bohr, R. Haubner, B. Lux: J. Refract. Hard Met. **14**, 145 (1996)
53. R. Paine, C. Narula: Chem. Rev. **90**, 73 (1991)
54. T.R. Anthony: Vacuum **41**, 1356 (1990)
55. J.C. Angus, A. Argoitia, R. Gat, Z. Li, M. Sunkara, L. Wang, Y. Wang: Phil. Trans. R. Soc. Lond. A **342**, 195 (1993)
56. V. Solozhenko: Diamond Relat. Mater. **4**, 1 (1994)
57. F. Corrigan, F. Bundy: J. Chem. Phys. **63**, 3812 (1975)
58. B. Singh, G. Nover, G. Will: J. Crys. Growth **152**, 143 (1995)
59. R.H. Wentorf: J. Chem. Phys. **34**, 809 (1961)
60. R.H. Wentorf: J. Chem. Phys. **36**, 1990 (1962)
61. L. Vel, G. Demazeau, J. Etourneau: Mat. Sci. Eng. B **10**, 149 (1991)
62. G. Demazeau: Diamond Relat. Mater. **2**, 197 (1993)
63. J. von der Gönna, H.J. Meurer, G. Nover, T. Peun, D. Schönbohm, G. Will: Mater. Lett. **33**, 321 (1998)
64. B. Lux: (1995), private communication
65. K.P. Loh, I. Sakaguchi, M. Nishitani-Gamo, T. Taniguchi, T. Ando: Phys. Rev. B **56**, 12 791 (1997)
66. T. Taniguchi: "Improvement of c-BN thin film adhesion: Theoretical considerations", in *Advanced Materials '98*, ed. by M. Kamo, T. Taniguchi, H. Yusa, H. Kanda, S. Koizumi, Y. Matsui, T. Aizawa, M. Kobayashi, K. Seki (NIRIM, Tsukuba, Japan, 1998), p. 59
67. J. Widany: (1996,1997), private communications

68. D. Bouchier, W. Möller: Surf. Coat. Technol. **51**, 190 (1992)
69. W.A. Yarbrough: J. Vac. Sci. Technol. **A9**, 1145 (1991)
70. J.E. Field: Inst. Phys. Conf. Ser. **75**, 181 (1986)
71. D. Marton, K.J. Boyd, J.W. Rabelais: Int. J. Mod. Phys. B **9**, 3527 (1995)
72. S. Reinke: (1996), "Modellierung der Dünnschichtabscheidung von kubischem Bornitrid", Ph.D. thesis, Kassel
73. C. Ronning: (1996), "Untersuchungen zum Wachstum und zur Dotierung diamantähnlicher Schichten, hergestellt über die Deposition massenseparierter Ionen", Ph.D. thesis, Konstanz
74. H. Vogel: *Gerthsen Physik* (Springer, Heidelberg, Berlin, 1995)
75. L. Bergmann, C. Schäfer: *Lehrbuch der Experimentalphysik*, Vol. I, Mechanik Akustik Wärme (Walter de Gruyter, Berlin, 1974), Chap. 8, Allgemeine Wellenlehre, p. 443
76. P.E. Pehrsson, F.G. Celii, J.E. Butler: "Chemical mechanisms of diamond CVD", in *Diamond Films and Coatings*, ed. by R.F. Davis (Noyes, Park Rigde, NJ, 1993)
77. C. Kittel: *Introduction to Solid State Physics* (Wiley, New York, 1996), Chap. 6
78. T.R. Anthony, W.F. Banholzer, J.F. Fleischer, L. Wei, P.K. Kuo, R.L. Thomas, R.W. Pryor: Phys. Rev. B **42**, 1104 (1990)
79. R. Berman: Phys. Rev. B **45**, 5726 (1992)
80. J.C. Phillips: *Bonds and Bands in Semiconductors* (Academic Press, New York, 1973)
81. J.C. Phillips: "Chemical models of energy bands", in *Handbook on Semiconductors*, ed. by T.S. Moss, Vol. 1 (Elsevier, Amsterdam, 1992), p. 47
82. E. Palik (Ed.): *Handbook of Optical Constants of Solids* (Academic Press, Orlando, 1985)
83. J.C. Angus, Y. Wang, M. Sunkara: Ann. Rev. Mater. Sci. **21**, 221 (1991)
84. P.K. Bachmann, R. Messier: Chemical & Engineering News p. 25 (1989)
85. P.K. Bachmann, W.v. Enckevort: Diamond Relat. Mater. **1**, 1021 (1992)
86. W.A. Yarbrough: J. Am. Ceram. Soc. **75**, 3179 (1992)
87. K.E. Spear, J.P. Dismukes (Eds.): *Synthetic Diamond: Emerging CVD Science and Technology* (Wiley, New York, 1994)
88. J.E. Butler, R.L. Woodin: Phil. Trans. R. Soc. Lond. A **342**, 209 (1993)
89. J. Wilson, W. Kulisch (Eds.): *Diamond Thin Films* (Akademie Verlag, Berlin, 1996)
90. J.C. Angus, F. Jansen: J. Vac. Sci. Technol. A **6**, 1778 (1988)
91. S. Reinke, R. Freudenstein, W. Kulisch: Surf. Coat. Technol. **98**, 263 (1998)
92. M. Weiler, S. Sattel, K. Jung, H. Ehrhardt, V. Veerasamy, J. Robertson: Appl. Phys. Lett. **64**, 2797 (1994)
93. H. List: "Bornitrid BN", in *Gmelin Handbuch der anorganischen Chemie*, ed. by K. Sommer, H.C. Buschbeck, Vol. 13, Borverbindungen, Part 1 (Springer, Heidelberg, Berlin, 1974)
94. A. Meller: "Boron Nitride BN", in *Gmelin Handbook of Inorganic Chemistry*, ed. by H.C. Buschbeck, K. Niendenzu, Vol. System No 13, Boron Compounds, 4th Supplement, Volume 3a (Springer, Heidelberg Berlin, 1988)
95. R. Locher, J. Wagner, F. Fuchs, M. Maier, P. Gonon, P. Koidl: Diamond Relat. Mater. **4**, 678 (1995)
96. S. Reinke, M. Kuhr, W. Kulisch, R. Kassing: Diamond Relat. Mater. **4**, 272 (1995)
97. W. Kulisch, S. Reinke: Diamond Films and Technology **7**, 205 (1997)
98. J. Widany, F. Weich, T. Köhler, D. Porezag, T. Frauenheim: Diamond Relat. Mater. **5**, 1031 (1996)

99. M. Kuhr: (1995), "Nukleations- und Wachstumsmechanismen kubischer Bornitrid-Filme", Ph.D. thesis, Kassel
100. D. Marton, K.J. Boyd, A.H. Al-Bayati, S.S. Todorov, J.W. Rabalais: Phys. Rev. Lett. **73**, 118 (1994)
101. J. Bulir, M. Jelinek, V. Vorlicek, J. Zemek, V. Perina: Thin Solid Films **292**, 318 (1997)
102. H. Sjöström, I. Ivanov, M. Johansson, I. Hultman, J.E. Sundgren, S.V. Hainsworth, T.F. Page, L.R. Wallenberg: Thin Solid Films **246**, 103 (1994)
103. H. Sjöström, S. Stafström, M. Boman, J.E. Sundgren: Phys. Rev. Lett. **75**, 1336 (1995)
104. S. Veprek, J. Weidmann, F. Glatz: J. Vac. Sci. Technol. A **13**, 2914 (1995)
105. L. Wan, R.F. Egerton: Thin Solid Films **279**, 34 (1996)
106. R. Beckmann: (1994), "Abscheidung von Diamant aus der Gasphase", Ph.D. thesis, Kassel
107. E. Oesterschulze, C. Mihalcea, W. Scholz, D. Albert, B. Sobisch, W. Kulisch: Appl. Phys. Lett. **70**, 435 (1997)
108. W. Kulisch, A. Malavé, G. Lippold, W. Scholz, C. Mihalcea, E. Oesterschulze: Diamond Relat. Mater. **6**, 906 (1997)
109. W. Scholz, D. Albert, A. Malavé, C. Mihalcea, W. Kulisch, E. Oesterschulze: "Fabrication of monolithic diamond probes for scanning probe microscopy applications", in *Micromachining and Imaging*, ed. by T. Michalske, M. Wendman (SPIE Proceedings Vol. 3009, 1997), p. 61
110. J.J. Hsieh: "Liquid–phase expitaxy", in *Handbook on Semiconductors*, ed. by T.S. Moss, S.P. Keller, Vol. 3 (North-Holland, Amsterdam, 1980), p. 415
111. J.D. Mackenzie: J. Non-Cryst. Solids **100**, 162 (1988)
112. B. Rother, J. Vetter: *Plasma-Beschichtungsverfahren und Hartstoffschichten* (Deutscher Verlag für Grundstoffindustrie, Leipzig, 1992)
113. R.F. Bunshah, C.V. Deshpandey: J. Vac. Sci. Technol. A **3**, 553 (1985)
114. W. Kulisch: Surf. Coat. Technol. **59**, 193 (1993)
115. W. Kulisch, H. tom Dieck: "Down Stream PE-MOCVD: Physikalisch-Technische Grundlagen der Abscheidung von metallischen und dielektrischen Schichten im Down-Stream-Plasma", in *Forschungsberichte des Bundesministeriums für Forschung und Technologie* (BMFT, 1992)
116. R.C. DeVries: Ann. Rev. Mater. Sci. **17**, 161 (1987)
117. J.C. Angus, C.C. Hayman: Science **241**, 913 (1988)
118. M. Sommer, K. Mui, F.W. Smith: Solid-State Commun. **69**, 775 (1989)
119. M. Sommer, F.W. Smith: High Temperature Science **27**, 173 (1990)
120. W. Banholzer: Surf. Coat. Technol. **53**, 1 (1992)
121. W.A. Yarbrough, R. Messier: Science **247**, 688 (1990)
122. W.A. Yarbrough: Diamond Films and Technologie **1**, 165 (1992)
123. P. Badziag, W.S. Verwoerd, W.P. Ellis, N.R. Greiner: Nature **343**, 244 (1990)
124. P.K. Bachmann, D. Leers, H. Lydtin: Diamond Relat. Mater. **1**, 12 (1991)
125. A. Badzian, T. Badzian, S.T. Lee: "Synthesis of diamond from methane and nitrogen mixture", in *Proceedings of the Third International Symposium on Diamond Materials*, ed. by J.P. Dismukes, K.V. Ravi, K.E. Spear, B. Lux, N. Setaka (The Electrochemical Society Proc. Vol. 93–17, Pennington, NJ, 1993), p. 378
126. T. Vandevelde, M. Nesladek, C. Quaeyhaegens, L. Stals: Thin Solid Films **308/309**, 167 (1997)
127. N.J. Komplin, J.B. Bai, C.J. Chu, J.L. Margrave, R.H. Hauge: "Chlorine–assisted chemical vapor deposition of diamond films at low temperatures", in *Proceedings of the Third International Symposium on Diamond Materials*, ed.

by J.P. Dismukes, K.V. Ravi, K.E. Spear, B. Lux, N. Setaka (The Electrochemical Society Proc. Vol. 93–17, Pennington, NJ, 1993), p. 385
128. F.C.N. Hong, G.T. Liang, J.J. Wu, D. Chang, J.C. Hsieh: Appl. Phys. Lett. **63**, 3149 (1993)
129. I. Schmidt, F. Hentschel, C. Benndorf: Diamond Relat. Mater. **5**, 1318 (1996)
130. P.K. Bachmann, H.J. Hagemann, H. Lade, D. Leers, D.U. Wiechert, H. Wilson, D. Fournier, K. Plamann: Diamond Relat. Mater. **4**, 820 (1995)
131. R. Beckmann, W. Kulisch, H. Frenck, R. Kassing: "Influence of gas phase parameters on deposition kinetics and morphology of thin diamond films by HFCVD technique at low temperatures", in *8th International Conference on Ion and Plasma Assisted Techniques (IPAT 91), Brussels*, ed. by N.E.W. Hartley (CEP Consultants Ltd, Edinburgh, May 1991), p. 64
132. R. Beckmann, W. Kulisch, H. Frenck, R. Kassing: Diamond Relat. Mater. **1**, 164 (1992)
133. R. Beckmann, W. Kulisch, H. Frenck, R. Kassing: "Influence of gas phase parameters on deposition kinetics and morphology of thin diamond films", in *Applied Diamond Conference (ADC 91), Auburn, Alabama*, ed. by Y. Tzeng, M. Yoshikawa, M. Muranaka, A. Feldmann (Elsevier, Amsterdam, 1991), p. 543
134. R. Beckmann, S. Reinke, M. Kuhr, W. Kulisch, R. Kassing: Surf. Coat. Technol. **60**, 506 (1993)
135. R. Beckmann, B. Sobisch, W. Kulisch: "Gas phase investigations during diamond deposisition with HFCVD and MW–PECVD", in *Proceedings of the Third International Symposium on Diamond Materials*, ed. by J. Dismukes, K. Ravi, K. Spear, B. Lux, N. Setaka (The Electrochemical Society Proc. Vol. 93–17, Pennington, NJ, 1993), p. 1025
136. R. Beckmann, B. Sobisch, W. Kulisch: (May 1994), "Basic gas phase mechanisms in diamond deposition from C/H/O mixtures", second International Symposium on Diamond Films (ISDF 2), Minsk, Belarus
137. R. Beckmann, B. Sobisch, W. Kulisch: Diamond Relat. Mater. **4**, 256 (1995)
138. N.A. Prijaya, J.C. Angus, P.K. Bachmann: Diamond Relat. Mater. **3**, 129 (1993)
139. C. Benndorf, I. Schmidt, P. Joeris: phys. stat. sol. (a) **154**, 5 (1996)
140. P. Joeris, C. Benndorf, S. Bohr: J. Appl. Phys. **71**, 4638 (1992)
141. D.K. Milne, P. John, I.C. Drummond, P.C. Roberts, M. Jubber, J.I.B. Wilson: (1994), "Emission spectroscopy measurements of diamond forming microwave plasmas", Hariot–Watt University, Edingburgh, unpublished results
142. W.D. Cassidy, P.W. Morrison, J.C. Angus: "Growth rates and quality of diamond grown by hot–filament assisted chemical vapor deposition", in *Diamond Materials IV*, ed. by P.K. Ravi, J.P. Dismukes (The Electrochemical Society Proc. Vol. 95–4, Pennington, NJ, 1995), p. 73
143. E.A. Evans, J.C. Angus: Diamond Relat. Mater. **5**, 200 (1996)
144. Y. Saito, K. Sato, H. Tanaka, K. Fujita, S. Matuda: J. Mater. Sci. **23**, 842 (1988)
145. Y. Liou, A. Inspektor, R. Weimer, R. Messier: Appl. Phys. Lett. **55**, 631 (1989)
146. C.F. Chen, S.H. Chen, H.W. Ko, S.E. Hsu: Diamond Relat. Mater. **3**, 443 (1994)
147. J. Stiegler, T. Lang, M. Nygard-Ferguson, Y. van Kaenel, E. Blank: Diamond Relat. Mater. **5**, 226 (1996)
148. Y.J. Baik, K.Y. Eun: "Growth shape of diamond in C–H–O system", in *Application of Diamond Films and Related Materials*, ed. by Y. Tzeng, M. Yoshikawa, M. Murakawa, A. Feldman (Elsevier, Amsterdam, 1991), p. 521

149. P. Bou, J.C. Boettner, L. Vanderbulcke: Jpn. J. Appl. Phys. **31**, 1505 (1992)
150. W.L. Hsu: J. Appl. Phys. **72**, 3102 (1992)
151. L. Okeke, H. Störi: Plasma Chem. Plasma Process. **11**, 489 (1991)
152. W. Kulisch: (1996), unpublished results
153. J. Wei, H. Kawarada, J. Suzuki, A. Hiraki: Jpn. J. Appl. Phys. **29**, L1483 (1990)
154. R.K. Singh, D. Gilbert, R. Tellshow, P.H. Holloway, R. Ochoa, J.H. Simmons, R. Koba: Appl. Phys. Lett. **61**, 2863 (1992)
155. R.A. Rudder, G.C. Hudson, J.B. Posthill, R.E. Thomas, R.C. Hendry, D.P. Malta, R.J. Markunas: Appl. Phys. Lett. **60**, 329 (1992)
156. J.C. Angus: (1994), private communication
157. W.D. Cassidy, E.A. Evans, Y. Wang, J.C. Angus, P.K. Bachmann, H.J. Hagemann, D. Leers, D.U. Wiechert: (1994), "Diamond growth rates and quality: Dependence on gas phase composition", paper presented at the Materials Research Society Meeting, San Francisco, California, April 4–8
158. J.C. Angus, W.D. Cassidy, L. Wang, E. Evans, C.S. Kovach: Mat. Res. Soc. Symp. Proc. **383**, 45 (1995)
159. D.G. Goodwin: J. Appl. Phys. **74**, 6895 (1993)
160. R. Kröger, L. Schäfer, C.P. Klages, R. Six: phys. stat. sol. (a) **154**, 33 (1996)
161. M.A. Kuzcmarski, P.A. Washlock, J.C. Angus: "Computer modeling of a hot filament diamond deposition reactor", in *Application of Diamond Films and Related Materials*, ed. by Y. Tzeng, M. Yoshikawa, M. Murakawa, A. Feldman (Elsevier, Amsterdam, 1991), p. 591
162. D.G. Goodwin: J. Appl. Phys. **74**, 6888 (1993)
163. M.A. Childs, K.L. Mennigen, H. Toyoda, Y. Ueda, L.W. Anderson, J.E. Lawler: Physics Letters A **194**, 119 (1994)
164. K.E. Spear, M. Frenklach: "Mechanisms of nucleation and growth of CVD diamond", in *Proceedings of the First International Symposium on Diamond and Diamond–Like Films, Los Angeles, May 1989*, ed. by J.P. Dismukes, A.J. Purdes, B.J. Meyerson, T.D. Moustakas, K.E. Spear, K.V. Ravi, M. Yoder (The Electrochemical Society, Pennington, NJ, 1989), p. 122
165. V.P. Varnin, I.G. Teremetskaya, D.V. Fedoseev, B.V. Deryagin: Sov. Phys. Dokl. **29**, 419 (1984)
166. A.R. Badzian, P.K. Bachmann, T. Hartnett, T. Badzian, R. Messier: "Diamond thin films prepared by plasma chemical vapor deposition processes", in *Amorphous Hydrogenated carbon films*, ed. by P. Koidl, P. Oelhafen (Les Editions de Physique, Paris, 1987), p. 63
167. E. Vietzke, V. Philipps, K. Flaskamp, P. Koidl, C. Wild: Surf. Coat. Technol. **47**, 156 (1991)
168. B.B. Pate: Surface Sci. **165**, 83 (1986)
169. T. Köhler, M. Sternberg, D. Porezag, T. Frauenheim: phys. stat. sol. (a) **154**, 69 (1996)
170. D. Albert: (1995), "Untersuchung des Einflusses von Sauerstoff auf die makroskopischen Schichteigenschaften von Diamant bei der Hot Filament Chemical Vapour Deposition", Master's thesis, Kassel
171. Y. Wang, J.C. Angus: "Microbalance studies of the kinetics of diamond growth", in *Proceedings of the Third International Symposium on Diamond Materials*, ed. by J.P. Dismukes, K.V. Ravi, K.E. Spear, B. Lux, N. Setaka (The Electrochemical Society Proc. Vol. 93–17, Pennington, NJ, 1993), p. 249
172. E. Kondoh, T. Ohta, T. Mitomo, K. Ohtsuka: Appl. Phys. Lett. **59**, 488 (1991)
173. K.A. Snail, C.M. Marks: Appl. Phys. Lett. **60**, 3135 (1992)

174. K.E. Spear, M. Frenklach: "Mechanisms for CVD diamond growth", in *Synthetic Diamond: Emerging CVD Science and Technology*, ed. by K. Spear, J. Dismukes (Wiley, New York, 1994), p. 243
175. M. Loh, M. Capelli: Diamond Relat. Mater. **2**, 454 (1993)
176. M. Frenklach, S. Skokov, B. Weiner: "On the role of surface diffusion in diamond growth", in *Diamond Materials IV*, ed. by P. Ravi, J. Dismukes (The Electrochemical Society Proc. Vol. 95–4, Pennington, NJ, 1995), p. 1
177. S.J. Harris, A.M. Weiner: Appl. Phys. Lett. **55**, 2179 (1989)
178. C. Wild, R. Kohl, N. Herres, W. Müller-Sebert, P. Koidl: Diamond Relat. Mater. **3**, 373 (1994)
179. C. Wild, N. Herres, P. Koidl: J. Appl. Phys. **68**, 973 (1990)
180. E.D. Specht, R.E. Clausing, L. Heatherly: J. Mater. Res. **5**, 2352 (1990)
181. R.E. Clausing, L. Heatherly, L.L. Horton, E.D. Specht, G.M. Begun, Z.L. Wang: Diamond Relat. Mater. **1**, 411 (1992)
182. C. Wild, P. Koidl, W. Müller-Sebert, H. Walcher, R. Kohl, N. Herres, R. Locher, R. Samlenski, R. Brenn: Diamond Relat. Mater. **2**, 158 (1993)
183. A. van der Drift: Philips Res. Rep. **22**, 267 (1967)
184. C.J. Chu, H. Hauge, J.L. Margrave, M.P. D'Evelyn: Appl. Phys. Lett. **61**, 1393 (1992)
185. H. Maeda, K. Ohtsubo, N. Ohya, K. Kusakabe, S. Morooka: J. Mater. Res. **10**, 3115 (1995)
186. R. Locher, C. Wild, N. Herres, D. Bohr, P. Koidl: Appl. Phys. Lett. **65**, 34 (1994)
187. S. Jin, T.D. Moustakas: Appl. Phys. Lett. **65**, 403 (1994)
188. J. Cifre, J. Puigdollers, M.C. Polo, J. Esteve: Diamond Relat. Mater. **3**, 628 (1994)
189. J.P. Gavigan: Diamond Relat. Mater. **1**, 1007 (1992). Procceedings of the Workshop on European Diamond Technology for the 90s, Brussels, Belgium, December 1991
190. J. Busch, J. Dismukes: "Economics of CVD diamond", in *Proceedings of the Third International Symposium on Diamond Materials*, ed. by J.P. Dismukes, K.V. Ravi, K.E. Spear, B. Lux, N. Setaka (The Electrochemical Society Proc. Vol. 93–17, Pennington, NJ, 1993), p. 880
191. S. Reinke, M. Kuhr, R. Beckmann, W. Kulisch, R. Kassing: "A sputter model for ion induced c–BN growth", in *Proceedings of the Third International Symposium on Diamond Materials*, ed. by J. Dismukes, K. Ravi, K. Spear, B. Lux, N. Setaka (The Electrochemical Society Proc. Vol. 93–17, Pennington, NJ, 1993), p. 283
192. S. Reinke, M. Kuhr, W. Kulisch: Diamond Relat. Mater. **3**, 341 (1994)
193. H. Saitoh, K. Yoshida, W. Yarbrough: J. Mater. Res. **8**, 8 (1993)
194. H. Feldermann, R. Merk, H. Hofsäss, C. Ronning, T. Zhaleva, R.F. Davis: (1998), "Laser approaches for deposition of carbon nitride films — chemical vapour deposition and ablation", paper presented at the 9th European Conference on Diamond, Diamond–Like and Related Material (DIAMOND'98), Crete; submitted to *Diamond Relat. Mater.*
195. C. Taylor II, R. Clarke: (1995), paper presented at the 6th European Conference on Diamond, Diamond-like and Related Materials (DIAMOND FILMS'95), Barcelona, September 1995.
196. P. Mirkarimi, K. McCarty, D. Medlin, T. Wolfer, T. Friedmann, E. Klaus, G. Cardinale, D. Howitt: J. Mater. Res. **9**, 2925 (1994)
197. A. Baranyai: (1998), "Herstellung von Hartstoffschichten im System B/C/N mittels plasmaunterstützter CVD und Charakterisierung ihrer Struktur und mechanischen Eigenschaften", Ph.D. thesis, Kassel

198. L. Hackenberger, L. Pilione, R. Messier, G. Lamaze: J. Vac. Sci. Technol. A **12**, 1569 (1994)
199. M. Kuhr, R. Freudenstein, S. Reinke, W. Kulisch: "Secondary growth mechanisms during the deposition of c–BN thin films", in *Diamond Materials IV*, ed. by P. Ravi, J. Dismukes (The Electrochemical Society Proc. Vol. 95–4, Pennington, NJ, 1995), p. 365
200. H. Lüthje, K. Bewilogua, S. Daaud, M. Johansson, L. Hultman: Thin Solid Films **247**, 40 (1995)
201. M. Kuhr, R. Freudenstein, S. Reinke, W. Kulisch, G. Dollinger, A. Bergmaier: Journal of Chemical Vapor Deposition **3**, 259 (1996)
202. M. Kuhr, R. Freudenstein, S. Reinke, W. Kulisch, G. Dollinger, A. Bergmaier: Diamond Relat. Mater. **5**, 984 (1996)
203. R. Freudenstein: (1997), "Untersuchungen zum Wachstum und zur Nukleation von c-BN–Schichten", Master's thesis, Kassel
204. T. Yoshida: Diamond Relat. Mater. **5**, 501 (1996)
205. R. Freudenstein, S. Reinke, W. Kulisch: Surf. Coat. Technol. **98**, 270 (1998)
206. M. Kuhr, S. Reinke, W. Kulisch: Surf. Coat. Technol. **74/75**, 712 (1995)
207. T. Ichiki, T. Momose, T. Yoshida: J. Appl. Phys. **75**, 1330 (1994)
208. A. Weber, U. Bringmann, R. Nikulski, C. Klages: Surf. Coat. Technol. **60**, 493 (1993)
209. W. Dworschak, K. Jung, H. Erhardt: Thin Solid Films **254**, 65 (1995)
210. N. Tanabe, T. Hayashi, M. Iwaki: Diamond Relat. Mater. **1**, 883 (1992)
211. M. Sueda, T. Kobayashi, H. Tsukamoto, T. Rokkaku, S. Morimoto, Y. Fukaya, N. Yamashita, T. Wada: Thin Solid Films **228**, 97 (1993)
212. O. Tsuda, Y. Yamada, Y. Tatebayashi, T. Yoshida: "Preparation of cubic boron nitride films by rf bias sputtering", in *12th International Symposium on Plasma Chemistry (ISPC95), Minneapolis, Minnesota, USA*, ed. by J. Heberlein, D. Ernie, J. Roberts, Vol. 4 (IUPAC, August 1995), p. 2041
213. S. Ulrich, J. Scherer, I. Schwan, I. Barzen, K. Jung, H. Ehrhardt: Diamond Relat. Mater. **4**, 289 (1995)
214. S. Reinke, M. Kuhr, W. Kulisch: Diamond Relat. Mater. **5**, 508 (1996)
215. S. Reinke, W. Kulisch: Surf. Coat. Technol. **98**, 23 (1998)
216. W. Kulisch, S. Reinke, R. Freudenstein: "Improvement of c-BN thin film adhesion: Theoretical considerations", in *Advanced Materials '98*, ed. by M. Kamo, T. Taniguchi, H. Yusa, H. Kanda, S. Koizumi, Y. Matsui, T. Aizawa, M. Kobayashi, K. Seki (NIRIM, Tsukuba, Japan, 1998), p. 63
217. R. Freudenstein, S. Reinke, W. Kulisch, R. Fischer, J. Zweck, A. Bergmaier, G. Dollinger: Mat. Sci. For. **287/88**, 259 (1998)
218. H. Hofsäss, C. Ronning, U. Griesmeier, M. Gross, S. Reinke, M. Kuhr: Appl. Phys. Lett. **67**, 46 (1995)
219. H. Hofsäss, C. Ronning, U. Griesmeier, M. Gross, S. Reinke, M. Kuhr, J. Zweck, R. Fischer: Nucl. Instr. and Meth. B **110**, 153 (1995)
220. B.N. Chapman: *Glow discharge processes* (Wiley, New York, 1982)
221. M. Komuna: *Film deposition by plasma techniques*, Springer Ser. Atoms and Plasmas, Vol. 10 (Springer, Berlin, 1992)
222. F. Seitz, J. Köhler: Solid State Phys. **2**, 305 (1956)
223. J. Robertson: Diamond Relat. Mater. **5**, 519 (1996)
224. C. Weissmantel, K. Bewilogua, D. Dietrich, H. Erler, H. Hinneberg, S. Klose, W. Nowick, G. Reisse: Thin Solid Films **72**, 19 (1980)
225. J. Robertson: Diamond Relat. Mater. **2**, 984 (1993)
226. C. Davis: Thin Solid Films **226**, 30 (1993)
227. H. Hofsäss, H. Feldermann, R. Merk, M. Sebastian, C. Ronning: Appl. Phys. A **66**, 153 (1999)

228. Y. Lifshitz, S. Kasi, J. Rabalais, W. Eckstein: Phys. Rev. B **41**, 10 468 (1990)
229. W. Dworschak: (1994), "Untersuchungen zum Wachstumsmechanismus von kubischen Bornitrid-Schichten", Ph.D. thesis, University of Kaiserslautern, Germany
230. J. Robertson: "Deposition of diamond–like carbon, cubic boron nitride and bias–enhanced nucleation of diamond as a subplantation process", in *Diamond Materials IV*, ed. by P. Ravi, J. Dismukes (The Electrochemical Society Proc. Vol. 95–4, Pennington, NJ, 1995)
231. D. McKenzie, W. McFall, W. Sainty, C. Davis, R. Collins: Diamond Relat. Mater. **2**, 970 (1993)
232. D. McKenzie, W. McFall, S. Reich, B. James, I. Falconer, R. Boswell, H. Persing, A. Perry, A. Durandet: Surf. Coat. Technol. **78**, 255 (1996)
233. F. Richter: "Cubic boron nitride and carbon nitride films: Recent developments", in *Diamond Materials IV*, ed. by P. Ravi, J. Dismukes (The Electrochemical Society Proc. Vol. 95–4, Pennington, NJ, 1995), p. 347
234. S. Reinke, M. Kuhr, W. Kulisch: Surf. Coat. Technol. **74/75**, 723 (1995)
235. M. Djouadi, D. Bouchier, P. Möller, G. Sené: "Effect of noble gas ions on the synthesis of cBN by ion beam assisted deposition", in *9th International Colloquium on Plasma Processes (CIP93)*Antibes (1993)
236. H. Hofsäss, H. Feldermann, M. Sebastian, C. Ronning: Phys. Rev. B **55**, 13 230 (1997)
237. J. Ye, U. Rothhaar, H. Oechsner: Surf. Coat. Technol. **105**, 159 (1998)
238. P. Sigmund: Phys. Rev. **184**, 383 (1969)
239. S. Ulrich: (1994), private communication
240. S. Matson-Kidner: (1994), "Ion assisted sputter deposition and structural characterization of cubic boron nitride", Ph.D. thesis, The University of Michigan, USA
241. Y. Osaka, A. Chayahara, H. Yokoyama, T. Hamada, T. Imura, M. Fujisawa: Mat. Sci. For. **54/55**, 277 (1990)
242. S. Westermeyr, M. Haag, J. Ye, H. Oechsner: (1996), "Characterization of boron nitride films and of their behaviour under ion irradiation with Auger electron spectroscopy (AES)", paper presented at ECASIA'96. Accepted for publication in *Surface and Interface Analysis*
243. J. Hahn, F. Richter, R. Pintaske, M. Röder, E. Schneider, T. Welzel: Surf. Coat. Technol. **92**, 129 (1997)
244. D. McKenzie, W. McFall, H. Smith, B. Higgins, R. Boswell, A. Durandet, B. James, I. Falconer: Nucl. Instr. and Meth. **106**, 90 (1995)
245. R. Freudenstein, S. Reinke, W. Kulisch: Diamond Relat. Mater. **6**, 584 (1997)
246. C. Uzan-Saguy, C. Cytermann, R. Brener, V. Richter, M. Shanaan, R. Kalish: Appl. Phys. Lett. **67**, 1194 (1995)
247. S. Yugo, T. Kimura, T. Muto: Vacuum **41**, 1364 (1990)
248. X. Jiang, W.J. Zhang, M. Paul, C.P. Klages: Appl. Phys. Lett. **68**, 1927 (1996)
249. D. Bouchier, G. Sené, M. Djouadi, P. Möller: Nucl. Instr. and Meth. B **89**, 369 (1994)
250. H. Naguib, R. Kelly: Rad. Eff. **25**, 1 (1975)
251. K. Osuch, W. Verwoerd: Surface Sci. **285**, 59 (1993)
252. S. Bohr, R. Haubner, B. Lux: Diamond Relat. Mater. **4**, 714 (1995)
253. S. Harris, G. Doll, D. Chance, A. Weiner: Appl. Phys. Lett. **67**, 2314 (1995)
254. S. Harris, A. Weiner, G. Doll, W. Meng: J. Mater. Res. **12**, 412 (1997)
255. F. Kiel: (1993), "Passivierung von Indiumphosphid im Niedertemperaturverfahren", Ph.D. thesis, University of Kassel
256. C. Schaffnit, L. Thomas, F. Rossi, R. Hugon, Y. Pauleau: Surf. Coat. Technol. **98**, 1262 (1998)

257. J. Ishikawa, Y. Takeiri, K. Ogawa, T. Takagi: J. Appl. Phys. **61**, 2509 (1987)
258. J. Koskinen: J. Appl. Phys. **63**, 2094 (1988)
259. H. Hofsäss, J. Biegel, C. Ronning, R. Downing, G. Lamaze: Mat. Res. Soc. Symp. Proc. **316**, 881 (1994)
260. Y. Lifshitz, G. Lempert, E. Grossmann, I. Avigal, C. Uzan-Saguy, R. Kalish, J. Kulik, D. Marton, J. Rabalais: Diamond Relat. Mater. **4**, 318 (1995)
261. D. McKenzie, D. Muller, B. Pailthorpe, Z. Wang, E. Kravtchinskaia, D. Segal, P. Lukins, P. Swift, P. Martin, G. Amaratunga, P. Gaskell, A. Saeed: Diamond Relat. Mater. **1**, 51 (1991)
262. D. McKenzie, Y. Yin, N. Marks, C. Davis, B. Pailthorpe, G. Amaratunga, V. Veerasamy: Diamond Relat. Mater. **3**, 353 (1994)
263. B. Rother, J. Siegel, I. Mühling, H. Fritsch, K. Breuer: Mat. Sci. Eng. A **140**, 780 (1991)
264. P. Fallon, V. Veerasamy, C. Davis, J. Robertson, G. Amaratunga, W. Milne, J. Koskinen: Phys. Rev. B **48**, 4777 (1993)
265. V. Veerasamy, G. Amaratunga, W. Milne, J. Robertson, P. Fallon: J. Non-Cryst. Solids **164–166**, 1111 (1993)
266. H. Scheibe, B. Schultrich: Thin Solid Films **246**, 92 (1994)
267. I. Schwan, S. Ulrich, H. Roth, H. Ehrhardt, S. Silva, J. Robertson, R. Samlenski, R. Brenn: J. Appl. Phys. **79**, 1416 (1996)
268. F. Richter, K. Bewilogua, I. Kupfer, I. Mühling, B. Rau, B. Rother, D. Schumacher: Thin Solid Films **212**, 245 (1992)
269. J. Cuomo, D. Pappas, R. Lossy, J. Doyle, J. Bruley: J. Vac. Sci. Technol. A **10**, 3414 (1992)
270. V. Puzikov, A. Semenov: Surf. Coat. Technol. **47**, 445 (1991)
271. W. Lau, I. Bello, L. Feng X. Huang, Q. Fuguang, Y. Zhenyu, R. Zhizhang: J. Appl. Phys. **70**, 5623 (1991)
272. Y. Lifshitz, G. Lempert, S. Rotter, I. Avigal, C. Uzan-Saguy, R. Kalish: Diamond Relat. Mater. **3**, 285 (1993)
273. I. Muraguchi, Y. Osaka: "Temperature effects on cubic BN films prepared by ECR plasma", in *Proceedings of the 6th Symposium on Plasma Science for Materials*, ed. by M. Kuzuya, A. Matsuda, Y. Joriike, Y. Moriyoshi, K. Takeda (Japan Society for the Promotion of Science, Commitee 153, Tokyo, Japan, 1993), p. 183
274. S. Ulrich, J. Schwan, W. Donner, H. Ehrhardt: Diamond Relat. Mater. **5**, 549 (1996)
275. T. Ikeda, T. Satou, H. Satoh: Surf. Coat. Technol. **50**, 33 (1991)
276. M. Weiler, R. Kleber, S. Sattel, K. Jung, H. Ehrhardt, G. Jungnickel, S. Deutschmann, U. Stephan, P. Blaudeck, T. Frauenheim: Diamond Relat. Mater. **3**, 245 (1994)
277. T. Frauenheim, G. Jungnickel, T. Köhler, U. Stephan: J. Non-Cryst. Solids **182**, 186 (1995)
278. D. McKenzie, D. Muller, B. Pailthorpe: Phys. Rev. Lett. **67**, 773 (1991)
279. M. Murakawa, S. Watanabe, S. Miyake: Diamond Films Technol. **1**, 55 (1991)
280. G. Sené: (1995), "Synthese de nitrure de bore cubique sous irradiation ionique: characterisation structurale et mechanismes de base", Ph.D. thesis, University of Paris-Sud
281. M.R. Wixom: J. Am. Ceram. Soc. **73**, 1973 (1990)
282. T. Sekine, H. Kanda, Y. Bando, M. Yokoyama, K. Hojou: J. Mater. Sci. Lett. **9**, 1376 (1990)
283. A.J. Stevens, T. Koga, C.B. Agee, M.J. Aziz, C.M. Lieber: J. Am. Chem. Soc. **118**, 10 900 (1996)
284. C.M. Sung, M. Sung: Mater. Chem. Phys. **43**, 1 (1996)

285. E.G. Wang: Progress in Material Science **41**, 247 (1997)
286. F. Fujimoto, K. Ogata: Jpn. J. Appl. Phys. **32**, L420 (1993)
287. H. Hofsäss, C. Ronning, U. Griesmeier, M. Gross: Mat. Res. Soc. Symp. Proc. **354**, 93 (1995)
288. F. Rossi, B. Andre, A. van Veen, P.E. Mijnarends, H. Schut, F. Labohm, M.P. Delplancke, H. Dunlop, E. Anger: Thin Solid Films **253**, 85 (1994)
289. M.Y. Chen, D. Li, X. Lin, V.P. Dravid, Y.W. Chung, M.S. Wong, W.D. Sproul: J. Vac. Sci. Technol. A **11**, 521 (1993)
290. K.M. Yu, M.L. Cohen, E.E. Haller, W.L. Hansen, A.Y. Liu, I.C. Wu: Phys. Rev. B **49**, 5034 (1994)
291. F.R. Weber, H. Oechsner: Surf. Coat. Technol. **74/75**, 704 (1995)
292. J.J. Cuomo, P.A. Leary, D. Yu, W. Reuter, M. Frisch: J. Vac. Sci. Technol. **16**, 299 (1979)
293. K.J. Boyd, D. Marton, S.S. Todorov, A.H. Al-Bayati, J. Kulik, J.W. Rabalais: J. Vac. Sci. Technol. A **13**, 2110 (1995)
294. A. Bousetta, M. Lu, A. Bensaoula, A. Schultz: Appl. Phys. Lett. **65**, 696 (1994)
295. C. Niu, Y.Z. Lu, C.M. Lieber: Science **261**, 334 (1993)
296. X.A. Zhao, C.W. Ong, Y.C. Tsang, Y.W. Wong, P.W. Chan, C.L. Choy: Appl. Phys. Lett. **66**, 2652 (1995)
297. A. Hoffman, I. Gouzman, R. Brener: Appl. Phys. Lett. **64**, 845 (1994)
298. O. Matsumoto, T. Kotaki, H. Shikano, K. Takemura, S. Tanaka: J. Electrochem. Soc. **141**, L16 (1994)
299. H. Hofsäss, C. Ronning, H. Feldermann, M. Sebastian: Mat. Res. Soc. Symp. Proc. **438** (1997)
300. K.G. Kreider, M.J. Tarlov, G.J. Gillen, G.E. Poirier, L.H. Robins, L.K. Ives, W.D. Bowers, R.B. Marinenko, D.T. Smith: J. Mater. Res. **10**, 3079 (1995)
301. Z.J. Zhang, S. Fan, J.L. Huang, C.M. Lieber: J. Electron. Mat. **25**, 57 (1996)
302. F. Weich, J. Widany, T. Frauenheim: Phys. Rev. Lett. **78**, 3326 (1997)
303. J. Mort, M.A. Machonkin, K. Okamura: Appl. Phys. Lett. **59**, 3148 (1991)
304. J.H. Edgar, Z.Y. Xie, D.N. Braski: Diamond Relat. Mater. **7**, 35 (1998)
305. J. Walker: Rep. Prog. Phys. **42**, 1605 (1979)
306. G.S. Woods: Proc. R. Soc. Lond. A **407**, 219 (1986)
307. H.X. Han, B.J. Feldman: Solid-State Commun. **65**, 921 (1988)
308. J. Kouvetakis, A. Bandari, M. Todd, B. Wilkens, N. Cave: Chem. Mater. **6**, 811 (1994)
309. Z.J. Zhang, S. Fan, C.M. Lieber: Appl. Phys. Lett. **66**, 3582 (1995)
310. W.T. Zheng, E. Sjöström, I. Ivanov, K.Z. Xing, E. Broitman, W.R. Salaneck, J.E. Greene, J.E. Sundgren: J. Vac. Sci. Technol. A **14**, 2696 (1996)
311. A.K. Sharma, P. Ayyub, M.S. Multani, K.P. Adhi, S.B. Ogale, M. Sunderaraman, D.D. Upadhyay, S. Banerjee: Appl. Phys. Lett. **69**, 3489 (1996)
312. C. Schaffnit, L. Thomas, R. Hugon, F. Rossi: Surf. Coat. Technol. **86/87**, 402 (1996)
313. A.F. Hollemann, E. Wiberg: *Lehrbuch der Anorganischen Chemie* (Walter de Gruyter, Berlin, 1985)
314. C. Chiang, H. Hollek, O. Meyer: Nucl. Instr. and Meth. B **91**, 692 (1994)
315. R. Kern, G. Le Lay, J.J. Metois: "Basic mechanisms in the early stages of epitaxy", in *Current Topics in Material Science*, ed. by E. Kaldis, Vol. 3 (North-Holland, Amsterdam, 1979), p. 131
316. J.A. Venables, G.L. Price: "Nucleation of thin films", in *Epitaxial Growth Part B*, ed. by J.W. Matthews (Academic Press, New York, 1975), p. 381
317. J.L. Robins: Appl. Sur. Sci. **33/34**, 379 (1988)
318. K. Reichelt: Vacuum **38**, 1083 (1988)

319. W.A.L. Lambrecht, C.H. Lee, B. Segall, J.C. Angus, Z. Li, M. Sunkara: Nature **364**, 607 (1993)
320. J.C. Angus, Z. Li, M. Sunkara, C.H. Lee, W.A.L. Lambrecht, B. Segall: "Diamond nucleation", in *Proceedings of the Third International Symposium on Diamond Materials*, ed. by J.P. Dismukes, K.V. Ravi, K.E. Spear, B. Lux, N. Setaka (The Electrochemical Society Proc. Vol. 93–17, Pennington, NJ, 1993), p. 128
321. D.N. Belton, S.J. Schmieg: Surface Sci. **233**, 131 (1990)
322. M.M. Waite, S.I. Shah: Appl. Phys. Lett. **60**, 2344 (1992)
323. S. Koizumi, T. Inuzuka: Jpn. J. Appl. Phys. **32**, 3920 (1993)
324. L. Wang, P. Pirouz, A. Argoitia, J.S. Ma, J.C. Angus: Appl. Phys. Lett. **63**, 1336 (1993)
325. T. Suzuki, M. Yagi, K. Shibuki, M. Hasemi: Appl. Phys. Lett. **65**, 540 (1994)
326. W. Zhu, P.C. Yang, J.T. Glass: Appl. Phys. Lett. **63**, 1640 (1993)
327. T. Suzuki, A. Argoitia: phys. stat. sol. (a) **154**, 239 (1996)
328. M. Schreck, B. Stritzker: phys. stat. sol. (a) **154**, 197 (1996)
329. T. Tomikawa, S. Shikata: Jpn. J. Appl. Phys. **32**, 3938 (1993)
330. Z. Li, L. Wang, T. Suzuki, A. Argoitia, P. Pirouz, J.C. Angus: J. Appl. Phys. **73**, 711 (1993)
331. G. Jungnickel, D. Porezag, T. Frauenheim, M.L. Heggie, W.R.L. Lambrecht, B. Segall, J.C. Angus: phys. stat. sol. (a) **154**, 109 (1996)
332. C.J. Chen, L. Chang, T.S. Lin, F.R. Chen: J. Mater. Res. **10**, 3041 (1995)
333. S. Iijima, Y. Aikawa, K. Baba: J. Mater. Res. **6**, 1491 (1991)
334. S. Iijima, Y. Aikawa, K. Baba: Appl. Phys. Lett. **57**, 2646 (1990)
335. S. Yugo, T. Kanai, T. Kimura, T. Muto: Appl. Phys. Lett. **58**, 1036 (1991)
336. F. Stubhan, M. Ferguson, H.J. Füsser, R.J. Behm: Appl. Phys. Lett. **66**, 1900 (1995)
337. J. Yang, Z. Lin, L. Wang, S. Jin, Z. Zhang: Appl. Phys. Lett. **65**, 3203 (1994)
338. J. Yang, Z. Lin, L. Wang, S. Jin, Z. Zhang: J. Phys. D **28**, 1153 (1995)
339. S.D. Wolter, J.T. Glass, B.R. Stoner: J. Appl. Phys. **77**, 5119 (1995)
340. H. Maeda, M. Irie, T. Hino, K. Kusakabe, S. Morooka: J. Mater. Res. **10**, 158 (1995)
341. K. Ohtsuka, K. Suzuki, A. Sawabe, T. Inuzuka: Jpn. J. Appl. Phys. **35**, L1072 (1996)
342. X. Jiang, C.P. Klages, R. Zachai, M. Hartweg, H.J. Füsser: Appl. Phys. Lett. **62**, 3438 (1993)
343. B.R. Stoner, J.T. Glass: Appl. Phys. Lett. **60**, 698 (1992)
344. W. Kulisch, L. Ackermann, B. Sobisch: phys. stat. sol. (a) **154**, 155 (1996)
345. X. Jiang, C.P. Klages: phys. stat. sol. (a) **154**, 175 (1996)
346. W. Kulisch, B. Sobisch, M. Kuhr, R. Beckmann: Diamond Relat. Mater. **4**, 401 (1995)
347. P. John, D.K. Milne, P.G. Roberts, M.G. Jubber, M. Liehr, J.I.B. Wilson: J. Mater. Res. **9**, 3083 (1994)
348. S.P. McGinnis, M.A. Kelly, S.B. Hagström: Appl. Phys. Lett. **66**, 3117 (1995)
349. R. Stöckel, K. Janischowsky, S. Rohmfeld, J. Ristein, M. Hundhausen, L. Ley: Diamond Relat. Mater. **5**, 321 (1996)
350. M.G. Jubber, D.K. Milne: phys. stat. sol. (a) **154**, 185 (1996)
351. X. Jiang, M. Paul, C.P. Klages: Diamond Relat. Mater. **5**, 251 (1996)
352. D.K. Milne, P.G. Roberts, P. John, M.G. Jubber, M. Liehr, J.I.B. Wilson: Diamond Relat. Mater. **4**, 394 (1995)
353. B.R. Stoner, G.H.M. Ma, D.S. Wolter, Y.C. Wang, R.F. Davis, J.T. Glass: Diamond Relat. Mater. **2**, 142 (1993)
354. K. Schiffmann, X. Jiang: Appl. Phys. A **59**, 17 (1994)

355. X. Jiang, C.L. Jia: Appl. Phys. Lett. **67**, 1197 (1995)
356. X. Jiang, C.L. Jia: Appl. Phys. Lett. **69**, 000 (1996)
357. M. Schreck, H. Roll, B. Stritzker: (1999), "SrTiO$_3$/Ir/diamond: A material combination for the preparation of single crystalline diamond films", to be published in *Appl. Phys. Lett.*
358. Y. Shigesato, R.E. Boekenhauer, B.W. Sheldon: Appl. Phys. Lett. **63**, 314 (1993)
359. S. Yugo, T. Kimura, T. Kanai: Diamond Relat. Mater. **2**, 328 (1993)
360. J. Robertson: Diamond Relat. Mater. **4**, 549 (1995)
361. J. Gerber, M. Weiler, O. Sohr, K. Jung, H. Ehrhardt: Diamond Relat. Mater. **3**, 506 (1994)
362. J. Gerber, S. Sattel, K. Jung, H. Ehrhardt, J. Robertson: Diamond Relat. Mater. **4**, 559 (1995)
363. L. Ackermann, R. Linnemann, M. Stopka, C. Mihalcea, S. M"unster, S. Werner, E. Oesterschulze, W. Kulisch: (1996), "Diamond film characterization by scanning probe microscopies", paper to be presented at the DIAMOND FILMS'96 Conference, Tours, France, September 1996
364. R. Beckmann, B. Sobisch, W. Kulisch, C. Rau: Diamond Relat. Mater. **3** (1994)
365. B.R. Stoner, G.H.M. Ma, D.S. Wolter, J.T. Glass: Phys. Rev. B **45**, 11 067 (1992)
366. J. Hahn: (1996), private communication
367. M. Schreck, T. Baur, B. Stritzker: Diamond Relat. Mater. **4**, 553 (1995)
368. B.W. Sheldon, Y. Shigesato, R.E. Boekenhauer, J. Rankin: "Diamond nucleation during bias–enhanced chemical vapor deposition", in *Proceedings of the Third International Symposium on Diamond Materials*, ed. by J.P. Dismukes, K.V. Ravi, K.E. Spear, B. Lux, N. Setaka (The Electrochemical Society Proc. Vol. 93–17, Pennington, NJ, 1993), p. 229
369. M.D. Whitfield, R.B. Jackman, D. Rodway, J.A. Savage, J.S. Foord: J. Appl. Phys. **80**, 3710 (1996)
370. S.P. McGinnis, M.A. Kelly, S.B. Hagström: "Mechanisms for the ion–assisted nucleation of diamond", in *Diamond Materials IV*, ed. by P.K. Ravi, J.P. Dismukes (The Electrochemical Society Proc. Vol. 95-4, Pennington, NJ, 1995), p. 73
371. X. Jiang, M. Paul, C.P. Klages, C.L. Jia: "Studies of heteroepitaxial nucleation and growth of diamond on silicon", in *Diamond Materials IV*, ed. by P.K. Ravi, J.P. Dismukes (The Electrochemical Society Proc. Vol. 95-4, Pennington, NJ, 1995), p. 50
372. S. Sattel, J. Gerber, H. Ehrhardt: phys. stat. sol. (a) **154**, 141 (1996)
373. A. Bergmaier: (1997), private communication
374. R. Chin, J.Y. Huand, Y.R. Shen, T.J. Chuang, H. Seki, M. Buck: Phys. Rev. B **45**, 1522 (1992)
375. B. Dischler, C. Wild, W. Müller-Sebert, P. Koidl: Physica B **185**, 217 (1993)
376. P. John, D.K. Milne, M.G. Jubber, J.I.B. Wilson, J.A. Savage: Diamond Relat. Mater. **3**, 488 (1994)
377. J. Gerber, S. Sattel, K. Jung, H. Ehrhardt: (1994), paper presented at the 8th CIMTEC Forum New Materials, Florence (Italy)
378. B. Sobisch, L. Ackermann, W. Kulisch: "Local characterization of bias-enhanced nucleation of diamond", paper presented at the 7th European Conference on Diamond, Diamond-like and Related Materials, (DIAMOND96), Tours
379. X. Jiang, K. Schiffmann, C.P. Klages: Phys. Rev. B **50**, 8402 (1994)

380. X. Jiang, K. Schiffmann, A. Westphal, C.P. Klages: Appl. Phys. Lett. **63**, 1203 (1993)
381. S.D. Wolter, B.R. Stoner, J.T. Glass, J.P. Ellis, D.S. Buhaenko, C.E. Jenkins, P. Southworth: Appl. Phys. Lett. **62**, 1215 (1993)
382. C.L. Jia, K. Urban, X. Jiang: Phys. Rev. B **52**, 5164 (1995)
383. M. Schreck, K.H. Thürer, R. Klarmann, B. Stritzker: Diamond Relat. Mater. **6** (1997)
384. W.J. Zhang, X. Jiang: Appl. Phys. Lett. **68**, 2195 (1996)
385. L.S. Yu, J.M.E. Harper, J.J. Cuomo, D.A. Smith: Appl. Phys. Lett. **47**, 932 (1985)
386. D. Kester, K. Ailey, R. Davis: J. Mater. Res. **8**, 1213 (1993)
387. M. Kuhr, S. Reinke, W. Kulisch: Diamond Relat. Mater. **4**, 375 (1995)
388. P. Mirkarimi, D. Medlin, K. McCarty, J. Barbour: Appl. Phys. Lett. **66**, 2813 (1995)
389. D. Kester, K. Ailey, R. Davis: Diamond Relat. Mater. **3**, 332 (1994)
390. S. Watanabe, S. Miyake, W. Zhou, Y. Ikuhara, T. Suzuki, M. Murakawa: Appl. Phys. Lett. **66**, 1478 (1995)
391. H. Yamashita, K. Kuroda, H. Saka, N. Yamashita, T. Watanabe, T. Wada: Thin Solid Films **253**, 72 (1994)
392. W. Zhou, Y. Ikuhara, T. Suzuki: Appl. Phys. Lett. **67**, 3551 (1995)
393. M. Röder, J. Hahn, U. Falke, S. Schulze, F. Richter, M. Hietschold: Mikrochim. Acta **125**, 000 (1996)

Index

β-Si$_3$N$_4$, 18, 40
β-C$_3$N$_4$ films
– working definition (lack of), 40

ab initio calculations, 18, 22, 23, 28, 40, 117
abrasive treatment, 140–142, 169
adhesion, 129, 133
– c-BN, 39, 42, 86, 166
adhesion energy, 126, 127, 129, 140
adhesion layers, 160
amorphization, 90, 103, 111, 115, 121
amorphous carbon, 18, 21, 35–37, 64, 73, 152, 154, 156
– classification, 36
– DLC, 18, 35, 37
– ta-C, 21, 26, 36, 37, 42
– ta-C:H, 38, 106
atomic hydrogen
– abstraction reactions, 46, 50, 57, 66–68
– and c-BN, 79, 104
– and ta-C, 111
– concentration, 48, 53, 58, 62, 67, 70, 71, 78, 146, 147
– creation of growth sites, 46, 65, 67, 103
– energy transport, 46, 67
– hydrogen termination, 46, 65, 66
– recombination, 50, 68
– regulation of gas phase composition, 46
– role during BEN, 150, 158, 159
– roles of, 46, 47, 65, 103, 120, 134
– selective etching, 46, 49, 64, 71, 73, 94, 102, 103
– stabilization of nuclei, 46, 50, 134, 138, 146
– surface stabilization, 46, 65, 103
– transport of, 61–63

B/C/N system, 18, 19, 21–27, 120, 121

Bachmann diagram, 52, 55–59
band gap, 22, 28, 32–34
bond energy, 1, 11, 16, 27, 30
bond length, 1, 8, 11, 15–18, 24, 27, 28, 30, 32–34, 40
bond, nature of, 1, 5, 7, 11, 12
boron carbide, 18, 23, 79
boron incorporation probability, 83, 97, 99, 101
boundary layer, 59, 60, 63
brittle materials, 5, 11, 15, 16
bulk modulus, 1, 22, 27, 28, 30, 33
– and hardness, 12–14, 16, 17
– Cohen's model, 17
– data, 18, 23
bulk properties, 2, 5, 10

c-BN deposition
– direct subplantation model, 101, 166
– dynamic stress model, 90–93, 96, 166
– indirect subplantation model, 90–91, 96, 102, 166
– quenching model, 89–90, 96, 166
– sputter model, 90, 93–99, 101, 102, 159, 166
– static stress model, 91–93, 96, 166
c-BN films
– working definition, 39, 79
carbon nitride compounds, 12, 18, 40, 112
– β-C$_3$N$_4$, 18, 21, 40
– C$_x$N$_y$ films, 40, 41, 113
– modifications, 40
carburization, 134, 141, 151, 155, 168, 169
collision cascade, 88–91, 109
competitive growth, 94, 99, 101, 163, 165
composite materials, 9–11
compressibility, 1, 2, 11, 12, 22
contamination
– c-BN, 38, 39, 79, 85

– diamond, 31, 38, 77, 118
convection, 59–63, 68
coordination number, 8, 17, 18, 22, 28
cracks, 5, 8, 11, 15
– critical crack length, 11, 15
– propagation of, 5, 8, 10, 11, 15
critical cluster size, 46, 126, 127, 129, 130, 133, 136
critical shear stress, 7, 15
crystallinity
– β-C_3N_4, 40, 41, 114–115
– c-BN, 26, 89, 90, 103
– diamond-like materials, 118, 121, 122
CVD
– c-BN growth (chemical), 79, 103, 105, 120
– c-BN growth (ion-assisted), 79, 85–87
– definition, 43
– diamond growth, 44

Debye temperature, 28, 30–31
defects
– and stress, 86, 92, 95, 109, 157
– cracks, 5, 8
– diamond films, 36, 69–72, 78
– dislocation spirals, 131
– dislocations, 5, 8
– Frenkel pairs, 91, 92
– generation, 91, 92, 95, 109
– interstitials, 90–92, 95, 109
– ion-induced, 88, 89, 91, 121, 157
– nucleation at, 131, 136, 137, 140, 141, 143, 168, 169
– phonon scattering, 31
– relaxation, 31, 89–93, 95, 102, 109–111, 118, 122
densification, 89, 91, 101, 109–111, 118, 121, 122, 165, 169, 171
density
– atomic, 12, 27, 28, 30–32
– electron, 32, 147
– high-density regions, 91
– mass, 11, 28, 30, 35
– nucleation, 103, 131, 133, 136, 140–143, 147, 149, 151, 155, 157–159
– of amorphous carbon films, 36
– of B/C/N films, 121
– of BN films, 38, 83, 89
– of CN films, 115, 116, 120
– of ta-C films, 37
– phonon, 110
– plasma, 83, 119, 149
deposition techniques, 43–44

– β-C_3N_4, 112–114
– c-BN, 79–80
– diamond, 44–47
– ta-C, 106
desorption, 131
– coefficient, 48, 50
– of boron, 93, 94, 97
– of carbon, 151
– of hydrogen, 67, 68, 73
diamond films
– working definition, 35
diamond-like materials
– common deposition process, 27, 120–123
– common nucleation process, 168–171
– definition, 18
– list of, 21
– material design, 27, 35
– properties, 27–35
– properties (table), 28
diffusion, 10, 23, 88, 91–93, 130, 134
– gas phase, 59–63, 68
– surface, 74, 130, 131
diffusion barrier, 134, 155
diffusion layer, 59
dislocations, 5, 8, 15, 131, 132
– interaction of, 8
– movement, 5, 8–10, 15
– multiplication, 8
– sources, 9
dispersion hardening, 9
ductile materials, 5, 8, 15

elastic deformation, 15
elastic properties, 1, 2, 4, 16
elastic strain, 9, 10
electronegativity, 16, 33
energy input, 101, 106, 110
– direct, 98, 99, 101, 106, 107, 109
– indirect, 98, 101, 102, 106, 110, 163
epitaxial growth, 132, 141–143, 152, 155, 158
evolutionary growth, 75, 94

fracture stress, 6, 11, 15
Frank–Read mechanisms, 8, 15
fullerenes, 22

gas phase processes, 43
– c-BN growth, 87–88
– diamond growth, 46–59, 63, 77
gas phase temperature, 44–48, 52, 53, 57–61, 67, 74, 76, 78, 79
grain boundaries, 9, 11, 133, 160, 165

grain boundary hardening, 9
graphitization, 73, 103, 111, 122
growth parameter, 75, 76
growth rates
- c-BN, 39, 83, 86, 97
- diamond, 36, 44–46, 48, 53, 56, 58–61, 63, 67, 69–73, 77

H6 carbon, 22
Hall–Petch equation, 9, 10
halogens
- c-BN growth, 104, 105, 120
- diamond growth, 51, 120
- diamond-like materials, 120
hard materials
- concepts, 5, 8, 9
- ideal, 5–8
- real, 8–11
- superhard, 18
hardness
- Berkovitch, 3
- Brinell, 3, 4
- correlations, 1, 5, 12–19
- definition, 1
- Knoop, 3, 4
- lack of absolute scale, 1, 4
- measurement, 2–5, 16
- micro-, 5
- Mohs scale, 2, 4, 12, 13, 18
- of thin films, 4, 5
- Rockwell, 3, 4
- theory of, 1
- ultramicro-, 5
- Vickers, 2–4, 18, 28
- Wooddell scale, 13
heteroepitaxy, 36, 132–136, 141, 142, 158
homoepitaxy, 140, 168, 169
HPHT processes, 109, 112
- β-C_3N_4, 112, 113, 117, 118
- c-BN, 25, 26, 40, 90, 92, 93, 160
- diamond, 23, 26, 40

ion angle of incidence
- c-BN, 80, 96–97, 108
- ta-C, 108
ion bombardment
- β-C_3N_4 growth, 114, 118
- amorphization, 90, 103, 111, 115, 121
- BEN, 103, 147, 148, 150, 151, 156, 158, 159
- c-BN growth, 80, 81, 85, 92, 94, 97, 99, 105, 156

- c-BN nucleation, 163, 165
- damage, 103, 157
- defect generation, 88, 91, 92, 95, 109, 121, 157
- densification, 101, 121, 165, 171
- diamond films, 103
- diamond-like materials, 121, 122
- effects, 80, 81, 88–89
- elastic collisions, 88
- implantation, 88
- inelastic collisions, 88
- knock-on effect, 88
- sputtering, 88
- ta-C growth, 106, 108, 111, 156
ion energy
- β-C_3N_4, 112–114
- average, 84, 87, 148
- BEN, 148, 149
- c-BN, 80, 82, 85, 95, 96, 99, 100, 107
- distribution, 148, 149
- ta-C, 107, 109
ion mass, 80
ion/neutral ratio
- F (definition), 80
- Φ^* (definition), 99
- F^* (definition), 83
- c-BN, 80, 82–84, 93–97, 99, 101, 107, 110
- ta-C, 106, 107, 109, 110
ionicity
- according to Cohen, 17, 28
- according to Pauling, 16, 28, 122
- according to Phillips, 122
- and bulk modulus, 17, 26
- and crystallinity, 26, 90, 103, 115, 118, 122
- and reactivity, 26
- of B–N bond, 26, 90, 95, 103, 122
- of C–N bond, 115, 118, 122
- of diamond-like materials, 27, 32, 34, 40, 122
island growth (Volmer–Weber growth), 129, 132, 133, 154
isotopes, 31

knock-on probability, 91, 101

layer growth (Frank–van der Merwe growth), 125, 129, 132, 133, 136
lonsdaleite, 24, 35, 38

melting point, 16, 28–30
misfit dislocations, 132

molecular-dynamic calculations, 26, 104, 105, 109, 115, 116, 139, 140, 164, 165
multilayer systems, 10
multiphase systems, 10

nanocrystallites, 134, 137
- β-C_3N_4, 21, 40, 112, 114, 115, 117, 119
- c-BN, 39, 86, 91, 117, 160, 162
- in amorphous matrix, 10
nanoindentation, 5
nanotubes, 11, 22
nucleation
- β-C_3N_4, 167–169
- BEN, 59, 133, 135, 136, 141–159, 169
- c-BN, 90, 94, 95, 98–101, 159–167, 169
- diamond, 134–136
- diamond at defects, 136, 137, 140, 141
- diamond on HOPG, 138–140, 156, 160, 163–165, 169
- diamond-like material, 133, 168–171
- ta-C, 167–169
nucleation (classical), 125–133, 168
- atomistic description, 125, 129–131
- influence of defects, 131
- nucleation modes, 128, 129, 131, 132
- statistical description, 125, 128–129
- thermodynamic description, 125–128
nucleation sequence
- BEN, 151–152, 154
- c-BN, 160, 162, 168

Peclet number, 59–60
penetration probability, 101, 109
phase diagram
- boron nitride, 25
- carbon, 23, 24, 44, 136
phase transition, 25, 89, 97, 101, 102, 121, 166, 168, 169
plastic deformation, 5, 8, 15, 16
plastic properties, 1, 2, 4
Poisson's ratio, 28, 30, 95, 157
preferential bond formation, 94, 95, 101
preferential sputtering, 114, 115, 122, 123
PVD
- c-BN growth, 79, 85–87
- definition, 43

quality
- c-BN, 26, 85, 93, 111
- diamond, 26, 39, 53, 57, 58, 61, 63, 67, 69–74, 77, 78, 103, 117, 145, 157, 158
diamond-like materials, 31

random covalent network, 36, 110
reduced models of diamond growth, 65, 67–73, 77
refractive index, 28, 33–34
resistivity, 28

selective etching
- c-BN, 104–105, 120
- diamond, 46, 64, 71, 73, 74, 94, 102, 103, 134
selective sputtering
- and crystal orientation, 158
- B/C/N films, 121
- c-BN, 93–95, 97, 99, 102, 165
shear modulus, 7, 16
silicon carbide, 18, 121, 122, 135, 136, 140, 141, 151, 152, 154–156
solution hardening, 9
sticking coefficient, 48, 93, 97, 143, 146
stoichiometry
- β-C_3N_4, 40, 41, 112–116, 118, 119
- binary/ternary materials, 122, 123
- c-BN, 26, 38, 39, 79, 82, 85
Stranski–Krastanov growth, 129
stress
- amorphous carbon films, 37
- BEN, 151–153, 156–158
- biaxial, 86, 92, 164
- c-BN films, 39, 42, 86, 89, 92, 95, 111, 166, 167
- h-BN nucleation layer, 163, 165
- instantaneous, 92
- static, 92
- ta-C films, 38, 108
substrate temperature
- β-C_3N_4, 112–115
- BEN, 150, 151, 157, 159
- c-BN, 80, 82, 83, 85, 88–90, 93, 97, 102, 108, 111
- diamond, 36, 44, 45, 47–49, 57, 62, 64, 65, 67, 71–76, 121
- diamond-like materials, 118, 121, 122
- effects, 80, 81, 90
- nucleation, 125, 130, 132
- ta-C, 107, 108, 110, 111
superhard
- definition, 18
superhard materials

- ideal, 17–19, 21
- multilayers, 10
- nanocomposites, 10–11

supersaturation, 125, 127, 128
- atomic hydrogen, 78
- carbon, 134, 138, 156, 168

surface energy, 6, 15, 125–128, 133, 135, 136, 140, 171
surface migration, 68, 76, 130, 132
surface mobility, 146
surface processes, 43
- c-BN growth, 80, 88
- diamond growth, 46, 57, 63–74, 77

surface reconstruction, 65, 104, 105
surface stabilization
- c-BN, 104, 105
- diamond, 46, 50, 65, 66, 74, 103, 120, 134

ta-BN, 26, 38
ta-C deposition
- direct subplantation model, 90, 91, 109, 110, 167
- quenching model, 90, 109
- static stress model, 90, 109

ta-C films
- working definition, 37

tensile strength, 1, 2, 5–8, 11
texture, 132
- amorphous carbon layers, 156
- c-BN films, 160
- c-BN nucleation layer, 160–165, 167, 169
- diamond films, 57, 74, 76, 78, 142, 158
- fiber texture, 74
- texture evolution, 75, 94

thermal conductivity, 28, 31–32, 36, 78, 90
thermal expansion coefficient, 28–30, 135
thermal spikes
- cylindrical, 90
- definition, 88
- densification, 91
- dimensions, 88, 90
- effects, 81, 89, 90, 109
- local temperature, 90, 109, 111
- relaxation processes, 89, 91, 93, 95, 109, 110
- temperature distribution, 88

transport processes, 43
- c-BN growth, 87–88
- diamond growth, 46, 59–63, 77

turbostratic BN, 38, 39, 163
turbostratic carbon nitride, 113

umklapp processes, 31–32

van der Drift model, 94
van der Driftmodel, 75
velocity of sound, 28, 30, 31
volumetric lattice energy, 12–13

whisker, 8
work hardening, 9

Young's modulus, 5, 7, 10, 15, 30
- and stress, 92, 157
- data, 28, 95, 157
- definition, 6, 29

Springer Tracts in Modern Physics

140 **Exclusive Production of Neutral Vector Mesons at the Electron-Proton Collider HERA**
By J. A. Crittenden 1997. 34 figs. VIII, 108 pages

141 **Disordered Alloys**
Diffusive Scattering and Monte Carlo Simulations
By W. Schweika 1998. 48 figs. X, 126 pages

142 **Phonon Raman Scattering in Semiconductors, Quantum Wells and Superlattices**
Basic Results and Applications
By T. Ruf 1998. 143 figs. VIII, 252 pages

143 **Femtosecond Real-Time Spectroscopy of Small Molecules and Clusters**
By E. Schreiber 1998. 131 figs. XII, 212 pages

144 **New Aspects of Electromagnetic and Acoustic Wave Diffusion**
By POAN Research Group 1998. 31 figs. IX, 117 pages

145 **Handbook of Feynman Path Integrals**
By C. Grosche and F. Steiner 1998. X, 449 pages

146 **Low-Energy Ion Irradiation of Solid Surfaces**
By H. Gnaser 1999. 93 figs. VIII, 293 pages

147 **Dispersion, Complex Analysis and Optical Spectroscopy**
By K.-E. Peiponen, E.M. Vartiainen, and T. Asakura 1999. 46 figs. VIII, 130 pages

148 **X-Ray Scattering from Soft-Matter Thin Films**
Materials Science and Basic Research
By M. Tolan 1999. 98 figs. IX, 197 pages

149 **High-Resolution X-Ray Scattering from Thin Films and Multilayers**
By V. Holý, U. Pietsch, and T. Baumbach 1999. 148 figs. XI, 256 pages

150 **QCD at HERA**
The Hadronic Final State in Deep Inelastic Scattering
By M. Kuhlen 1999. 99 figs. X, 172 pages

151 **Atomic Simulation of Electrooptic and Magnetooptic Oxide Materials**
By H. Donnerberg 1999. 45 figs. VIII, 205 pages

152 **Thermocapillary Convection in Models of Crystal Growth**
By H. Kuhlmann 1999. 101 figs. XVIII, 224 pages

153 **Neutral Kaons**
By R. Belušević 1999. 67 figs. XII, 183 pages

154 **Applied RHEED**
Reflection High-Energy Electron Diffraction During Crystal Growth
By W. Braun 1999. 150 figs. IX, 222 pages

155 **High-Temperature-Superconductor Thin Films at Microwave Frequencies**
By M. Hein 1999. 134 figs. XIV, 395 pages

156 **Growth Processes and Surface Phase Equilibria in Molecular Beam Epitaxy**
By N.N. Ledentsov 1999. 17 figs. VIII, 84 pages

157 **Deposition of Diamond-Like Superhard Materials**
By W. Kulisch 1999. 60 figs. X, 191 pages

Springer and the environment

At Springer we firmly believe that an international science publisher has a special obligation to the environment, and our corporate policies consistently reflect this conviction.

We also expect our business partners – paper mills, printers, packaging manufacturers, etc. – to commit themselves to using materials and production processes that do not harm the environment. The paper in this book is made from low- or no-chlorine pulp and is acid free, in conformance with international standards for paper permanency.